THEMES

MARLENE & ROBERT McCRACKEN

ILLUSTRATED BY DIANA COLQUHOUN
SPECIAL ART ACTIVITIES SECTION BY WYN DAVIES

Animals

PEGUIS PUBLISHERS LIMITED
462 HARGRAVE STREET
WINNIPEG, MANITOBA, CANADA
R3A 0X5

© 1985, 1988 Marlene J. McCracken and
Robert A. McCracken

All rights reserved

First Printing: 1985
Second Printing: 1986
Third Printing: 1987
Revised: 1988

Canadian Cataloguing in Publication Data

McCracken, Marlene J., 1932-
    Animals

    Rev. ed. —

    (Themes, ISSN 0838-3219)
    ISBN 0-920541-12-7

1. Language arts (Primary). 2. Activity programs in education. 3. Animals — Study and teaching (Primary). I. McCracken, Robert A., 1926-   II. Davies, Wyn, 1931-   III. Conquhoun, Diana.   IV. Title. V. Series.

LB1528.M334 1988   372.6'044   C88-098079-6

Printed and bound in Canada by
Hignell Printing Limited, Winnipeg

CONTENTS

THE THEME SERIES
A. Introduction v
B. What is a theme? vi
C. Organizing the day for thematic teaching. vi

ANIMALS THEME
A. Concept Development Through Picture Classification 1
B. Language Activities Using the Sorting Results 8
C. Classification 16
D. Developing Vocabulary and Sentence Structure 34
E. Writing Stories From Frames and Story Structures 41
F. Origin Stories 42
G. Using Poetry Within An Animal Theme 45
H. Books to Use As Models for Children's Own Writing 62
I. Chants and Songs to Use Within an Animal Theme 63

ART ACTIVITIES ON AN ANIMAL THEME
A. The Image Expressed in Two Dimensions 72
B. The Image Expressed in Three Dimensions 82

ACKNOWLEDGMENTS 84

BIBLIOGRAPHY 85

THE THEMES SERIES

A. INTRODUCTION

The *Themes Series* for the primary grades is based upon the theory and teaching techniques presented in *Reading Is Only the Tiger's Tail, Reading, Writing and Language, Spelling Through Phonics,* and *Stories, Songs, and Poetry to Teach Reading and Writing* by the same authors. The ideas and techniques are reviewed here.

Reading and writing are forms of language. Children acquire any language in any form if they are immersed in the language in that form, and are required to practice. Appropriate practice is the key to efficient learning.

Reading and writing are complementary skills which should be taught concurrently. The insights that children get through authorship encourage their reading, and vice versa.

Any language begins with experiences that provoke meanings deep inside the brain. It is only through a meaning that form may be understood. Skills must be learned through need in a meaningful context.

Reading and writing begin with concepts and contents. From the whole, we move to the smaller parts and then to the bits and pieces. The pieces become understandable only from the context of the whole. With print, the whole is the entire story or poem. We move from that to the chapter, the paragraph, the sentence, the phrase, the word, the syllable, and finally to the letters.

Skills grow from need as children are provoked to think about their world, to examine the ideas of authors, and to express their observations and reactions. Traditionally, we have presumed that children must begin with the pieces and "master" them before going on to the stories. The cliche, "First we learn to read, and then we read to learn," reflects this belief. We believe that you read to learn and as a result learn to read.

Writing requires four things. There must be **ideas** and **words** to express the ideas; these are the *what* of writing. Children must be able to **spell,** and use the common **structures** if they are to write successfully; these are the *how* of writing. The teacher must teach content and ideas, develop and present the vocabulary needed for the content, teach children how to spell and how to use the basic structures of written English, then demand that the children practice.

Reading can be divided into three parts. For these, we are indebted to Leland B. Jacobs, professor emeritus, Teachers College, Columbia University. (1) There is **reading to** children. This develops content, of course, but it also develops the melodies and structures of written English; it is the readiness for being able to read later by oneself. This is informal, or indirect teaching. (2) There is **reading with** children where the teacher directly works with print. This is formal, or direct teaching. (3) There is **reading by** the children. This is when they practice what has been taught.

B. WHAT IS A THEME?

A theme is an idea of concept, one that may be perceived because it recurs as the teacher teaches an idea or concept through many disciplines, and has the children explore the concept through all modalities. The term is used commonly in music to refer to a repeated melody played in several ways within a musical piece. Theme teaching is similar.

A theme may begin in somewhat rough form. The teacher reveals many facets; then the children work to discover more and to polish those which they already know. As the pupils' understanding becomes sophisticated, their language skills become more sophisticated through practice.

The teacher's job is threefold:
1. To decide which theme to teach,
2. To teach it as efficiently as possible, and
3. To demand that the pupils examine and work with the theme.

All basic language skills are required for theme teaching and they are developed and learned through meaningful practice. Theme work develops and extends a child's life and learning experiences cognitively and affectively. The learning is individualized as each child practices within his/her ability, interest, and understanding. The teacher assures growth by demanding ever increasing sophistication in skill practice and thinking.

Theme teaching is social, as is all natural language learning. Children and teacher work cooperatively to learn together and to assure that every child learns. Theme teaching focuses attention upon the content and the nitty-gritty skills. In primary grades there has been a tendency for skills to be the whole focus and goal. The teacher must maintain a balance between content and skill demands so that the children always are aware that *thought-giving* and *thought-getting* are the purposes for the skill learning. The children always write to record or communicate; they always read to acquire information. The skills are a natural result of practicing writing and reading. Children always read to learn, with the result that they learn to read.

C. ORGANIZING THE DAY FOR THEMATIC TEACHING

We begin each day by working orally with the entire class. We work to develop thinking about some part of the theme. We brainstorm, recording the ideas on the chalkboard, chanting to develop oral and written language structures, and classifying to develop thinking and further understanding of the ideas. This oral period is 45 minutes or more. The oral period is followed by children practicing what has been taught, usually writing for another hour or more. Small groups are taken out for instruction during this time.

Each day:

- We read to children.
- We read with children, directly instructing them in some comprehension techniques.

   This is often small group instruction, but seldom should it be grouped by ability.
- The children read silently.
- We sing and chant.
- The children write every day.

We have the children practice orally the theme content within the structures they will use when writing. Without the oral practice, many children are unable to work independently for an hour or more. Through their daily writing, children internalize all the nitty-gritty skills of language, the sound it represents, the use of capitals, and so on.

The day is integrated through the content, and the skills learned through practice. The teacher teaches, and demands that the children practice to the best of their abilities. The teacher is gentle, loving, but firm, telling the child that his/her best work is expected. This is a highly individualized program; the individualization comes through the practicing as children learn both content and skills.

ANIMALS THEME

INTRODUCTION

The **Animals Theme** focuses upon the lives, habits and appearances of animals. The theme probably fits best under the heading of science, but it branches into social studies, geography, and literature. The purpose of **Animals** is to increase the children's understanding of the world about them, to add to their knowledge of animal life, to develop the vocabulary peculiar to the study of animal life, and to begin to understand the interaction of ecological systems.

Reading and writing are taught as skills evolving from the theme. The skills are the necessary and inevitable results of thinking, communicating, and the recording of observations, information, and experiences. Thinking develops primarily from the classification of observations and facts. We join like bits and exclude unlike bits to form concepts, and we reform concepts as we learn more and think more. For this reason we begin animal study through classification, having children classify known animals. We add to the known throughout the theme. We read to children to introduce unfamiliar animals, they read about animals familiar and unfamiliar, and we provide activities through which the children refine and sophisticate their knowledge.

A. CONCEPT DEVELOPMENT THROUGH PICTURE CLASSIFICATION

This activity requires a set of animal picture cards of all types of animal life. The pictures need to be fairly detailed if in-depth work is to be done. A minimum of 150 animals will be needed. The children can help make this set by cutting and pasting pictures on cards.

1. **Children Sort Freely**

    Put children in groups of four. The groups should be random, more or less, and definitely not according to ability. Give each group of children at least twenty animal picture cards. Tell the children to sort the cards into two piles according to some rule. Have them figure out a rule and then sort the cards, keeping the rule a secret.

    When the cards are sorted the teacher tries to guess the rule. The teacher notes carefully how the children sort because this indicates what the children already know. The teacher can then build on known concepts to increase the children's familiarity with animals. All grade levels begin this way even though picture sorting may seem at first to be only a kindergarten or grade one activity.

2. **Children Sort to an Oral Rule**

    a) When it becomes obvious that some concept needs sophistication or introduction, the teacher sets rules. For example, if the concept of size needs to be developed with very young children she might begin as follows:

The teacher gives each child a two-inch paper square and asks them which animals could hide under it. The teacher draws their responses on the chalkboard. They chant from the chalkboard list. ('A bee is tiny. An ant is tiny. And so on.') Later the teacher draws the same animals on cards and hangs them on masking tape. Each day the class chants the list in sentences, learning the animal names and various ways to express the concept of **tiny**.

A worm is a tiny animal.
A lady bug is a tiny animal.
A bee is a tiny animal.
A spider is a tiny animal.

The worm is a little animal.
The lady bug is a little animal.
The bee is a little animal.
The spider is a little animal.

A worm is a small animal.
A lady bug is a small animal.
A bee is a small animal.
A spider is a small animal.

After several days of chanting give each child a tiny book, with four to eight blank pages. Three inches square or a three by five rectangle are good sizes. The children draw their favorite tiny animals, one to a page, and write or dictate to the teacher using one of the sentence patterns they have chanted. They might end their book with the smallest of the small saying,

And a _____ is the smallest animal I know.

The teacher may then ask the children to sort by putting all the animals that are bigger than they are in one pile. These large animals are placed on a list and words denoting 'large' are chanted.

An elephant is large.
A tiger is large.
A rhinoceros is large. (big, huge, enormous, monstrous)

The children make a big book with appropriate captions. With both the **tiny** and **big** books we work briefly with all children to observe the parts of each animal so that the drawings improve and are reasonable representations. We practice on small chalkboards. Children now work with all animals, placing them into three size categories:

Smaller than I am.
About the same size as I am.
Larger than I am.

The children can create books using several sentence patterns to express size. Any of the patterns used should be developed orally before expecting the children to use them in writing or dictation.

A _____ is smaller than I am.
A _____, a _____, and a _____ are smaller than I am.
I am bigger than a _____.
I am bigger than a _____ or a _____.
A _____ is smaller than a _____.
A _____ is smaller than a _____ or a _____.
A _____ is about the same size as a _____.
A _____ is about the same size as a small _____.

In kindergarten we expect no writing. We merely do the oral work and have the text chanted as part of the chalkboard or pocket chart presentation.

The various categories could be collated into class books, with each of the children listed as one of the authors.

Older children can research to discover the adult weights of the animals, sorting them into various weight categories. They can collect pictures of forty or more breeds of dogs and sort them into small, medium and large categories. The Kids' Dog Book is a marvelous source of information about dogs. Again, they may work with the adult weights to sort those that are smaller, about the same size, and larger than I am. They might work to rank order the breeds according to weight.

b) Similarly, the teacher may develop the following concepts by stating an oral rule and having children sort:

• Put all the animals that live in water in one pile.
Put all the animals that live on land in one pile.

Some animals will fit into both categories. This can be solved by using hula hoops or any large circles to create overlapping categories. See the illustration.

- Sort according to the number of legs.
- Sort by outside coverings: skin, hair, fur, feathers.
- Sort by color.
- Sort by whether the animal would be a good pet.
- Sort by whether the animal has been domesticated.

c) Use the poem, Dogs and Cats and Bears and Bats, by Jack Prelutsky.

DOGS AND CATS AND BEARS AND BATS

Mammals are a varied lot;
some are furry, some are not;
many come equipped with tails;
some have quills, a few have scales.

Some are large, and others small;
some are quick, while others crawl;
they prance on land, they swing from trees;
they're underground and in the seas.

Some have hooves, and some have paws;
some have fangs in snapping jaws;
some will snarl if you come near;
others quickly disappear.

Dogs and cats and bears and bats,
all are mammals, so are rats;
whales are mammals, camels too;
I'm a mammal . . . so are YOU!

Extract the classifications from the poem (furry, have tails, quills, and so on) from the poem and have the children sort to several or all of the classifications. (Note: In all of the classification activities children will make lots of mistakes. The emphasis should be upon classifying, socially arguing with peers, and gradually learning the right answers while practicing the concept.)

3) Children Sort From A Picture Without The Rule Being Stated.

a) The teacher selects one animal picture with some attribute she wishes to highlight: webbed feet, shell, bushy tail, hoofs, skinny tail, horns, mane, sharp teeth, pointed beak, scales, feathers, fins. She puts the picture in the top pocket of the pocket chart and gives each child three different animal pictures. She asks the children to find another picture that goes with the first picture. If the picture fits according to the teacher's rule, she has it placed in the top pocket. If the picture doesn't fit it is not accepted because, 'It doesn't fit my rule.' She may ask the child why it fits, to find out how the child was observing the likeness before saying, 'Yes, but that isn't my rule.'

If no child gets a correct match after four or five tries the teacher places a second card in the chart and challenges the children to find another. Once the rule has been discovered, all of the matching cards are placed in the pocket chart, and then the pictures are chanted to teach the category.

If webbed feet was the category the children would chant:

A _____ has webbed feet.
or
A _____, a _____, and a _____ all have webbed feet.

b) Once several categories have been developed, the children can make a graph to compare and contrast animal characteristics. Each group of children would get fifty animal cards. They would record the animal names on a graph with some animals being listed in several categories. For young children four categories might be enough, and with older children ten categories might be a maximum for the activity. Create a work sheet as follows:

webbed feet	hoofs	sharp teeth	claws	feathers	manes	horns

c) The children can then write their results. They might use the following pattern:

We classified fifty animals,
_____ had webbed feet, _____ had hoofs, and _____ had sharp teeth. _____ had claws, _____ had feathers, _____ had manes, and _____ had horns, _____ had claws and sharp teeth. _____ had webbed feet and feathers.

Continue until all the combinations have been enumerated. Numbers go into the blanks.

4. Sorting Pictures To Develop Concepts

a) Eating Habits

Supply a picture of a hoofed foot. Supply another of a clawed foot. Have the children sort into three groups: hoofs, claws, or neither. Ask the children what the animals in each classification eat, and develop this briefly. Study the teeth of each group and relate this to the eating habits. The children will probably note that hoofs are related to grazing animals, and molars or grinding teeth are also characteristic of the grazers or plant eaters. The clawed animals are likely to divide themselves into two groups, the meat catchers and the diggers who are differentiated by teeth.

Study the beaks of birds. If enough bird pictures are available, note the relationship to this and feeding habits. A beginning of beak study might be the curved beak of the eagle or hawk, and the straight beak of the robin.

b) Habitat

Supply pictures of a forest, grasslands, a pond, a jungle, the ocean, and the ocean shore. Have the children sort and make a large graph of habitat. If the eating habits have been studied, relate this to habitat.

Another sorting may be done by webbed feet. This will be mostly birds, and it will be the swimmers.

Use the poem 'Some Animals' by Robert McCracken. Place one or more stanzas in the pocket chart. Have the children decide which animals can replace the word *some*. Place the picture card of that animal on top of *some* and chant several variations of the poem.

SOME ANIMALS

Some dive in the oceans.
Some gambol on land.
Some soar through the blue sky.
Some sit on the strand.

Some gallop across prairies.
Some leap through the trees.
Some slither through marshlands.
Some creep underseas.

Some slide down the mudbanks.
Some stalk on the plain.
Some swim in the clear lakes.
Some walk in the rain.

Some often stay stone still.
Some inexorably stalk.
Some race through the grasslands.
But only one can talk.

Available as a poetry poster set. See bibliography.

c) **Survival**

Relate animals' outer coverings and colorations to habitat. Note those animals for which camouflage is important for survival.

Develop the concept of migration. Margaret Wise Brown's *Wheel on the Chimney* does this in a story about a stork. (Also available as a film and filmstrip from Weston Woods Studios.)

Classify the migrators. Those animals that do not migrate may be subclassified into those that hibernate, and those that stay awake during the winter.

d) **Birth**

Some animals are born from eggs; some are born alive. Ruth Heller's two books, *Animals Born Alive and Well* and *Chickens Aren't the Only Ones* are classics of verse and illustration. They should be read to children with the children then classifying pictures to develop the concepts.

e) **Movement**

Use the poem, "**How Creatures Move**," to introduce movement.

> The panther walks on padded paws,
> Squirrel leaps from limb to limb,
> A fly can crawl straight up a wall
> And seals can dive and swim.
> The snake can slither on the ground.
> The monkey hooks his tail and swings
> The robin hops to get around
> While eagles soar on outstretched wings.
> But boys and girls have much more fun.
> They leap and dance and walk and run.
> They jump and stroll. They skip and hop.
> They even climb a mountain top.
> (adapted by Robert McCracken from an anonymous poem)

Brainstorm for words that tell how animals move (walk, run, swim, slither, climb, fly, etc.). Refer to *Some Animals*. Try to get 20 or more words with older children (Scurry, scoot, scamper, gambol, and so on). Develop each movement with creative drama. Children use their forearms as the ground, and one hand as the animal to demonstrate each movement. Do the same things outdoors or in the gym using the whole body. Slither through the grass. Gambol across the field. Have children then sort pictures according to how the animals might move.

B. LANGUAGE ACTIVITIES USING THE SORTING RESULTS

1. Write Song Parodies

a) Children tell an animal's attributes using the melody and form of 'My Hat It Has Three Corners.' This can be made into a class book with each child contributing a page or more. Use the following examples as models:

My cat it has some whiskers,
Some whiskers has my cat.
And if he had no whiskers,
He would not be my cat.

My rooster has a comb,
A comb has my rooster.
And if he had no comb,
He would not be my rooster.

My horse it has a mane,
A mane has my horse,
And if she had no mane,
She would not be my horse.

My piggie has a snout,
A snout has my piggie.
And if it had no snout,
It would not be my piggie.

My bear he has some fur,
It covers all his skin.
And if he had no fur,
He'd be a bare bear.

Children can also write to "The Farmer in the Dell":

The squirrel has a tail.
The squirrel has a tail.
It keeps him warm at night.
The squirrel has a tail.

The horse has a mane.
The horse has a mane.
It grows upon his neck.
The horse has a mane.

b) The song 'Good-night Irene,' as adapted by Raffi, is a naming of habitat. Have the class make new verses:

Horses sleep in their stalls.
Owls sleep in trees.
Beaver sleep in their lodges.
And fishes sleep in the seas.

Chipmunks sleep in their burrows.
A bird sleeps in its nest.
A monkey sleeps in the tree tops.
And coyotes sleep in the West.

2. Write Book Parodies

a) Use *Where Do You Live?* by Robert and Marlene McCracken (Tiger Cub Books) to develop a writing pattern for the study of habitat. The book begins:

Where do you live?
Do you live in a ⎡dog house?⎤

No, no, no!
⎡Dogs⎤ live in a ⎡dog house.⎤

This can be built in the pocket chart with the words *dog house* and *Dogs* turned over to develop a writing frame. Work orally to develop several answers for some questions. For example, the question **Do you live in the ocean?** has dozens of answers. The teacher should develop specificity of home names and introduce the vocabulary of warren, lodge, colony, etc.

For beginning children the teacher may use pictures to name the habitat with children chanting the whole verse while adding an answer.

Class books could be made for each habitat. Every child writes and illustrates a page with an appropriate caption.

b) Use similarly *What Do You Have?, Some Dogs,* and *What Do You Do?* by Robert and Marlene McCracken (Tiger Cub Books.) Children may summarize their classifying modelling from *Some Dogs:*

　　Some ___snakes___ live in the ocean.
　　Some ___snakes___ don't.

The children then illustrate with sea snakes in one picture and land snakes in another. Or they might make a general statement as follows:

　　Some animals live in the ocean.
　　Some animals don't.

　　Some animals live in holes in the ground.
　　Some animals don't.

They then illustrate with several appropriate animals that live in the ocean or in the ground, and several that don't.

c) Make class Big Books for each habitat. Use the structure of *Brown Bear* by Bill Martin. Fern Finn in the Richmond, B.C. School District, worked with her grade one children in studying animal habitat. The children chose a habitat and worked in groups to create a book patterned after *Brown Bear*. They finished each poem with a line naming the habitat.

　　Coyote, coyote,
　　What do you see?
　　I see a mouse
　　Running across me.

　　Little mouse, Little mouse
　　What do you see?
　　I see a snake
　　Slithering toward me.

　　Long snake, long snake,
　　What do you see?
　　I see a butterfly
　　Flying above me.

> Butterfly, butterfly,
> What do you see?
> I see a deer
> Leaping at me.
>
> Big deer, big deer,
> What do you see?
> I see a pretty bird
> Flying above me.
>
> Pretty bird, pretty bird,
> What do you see?
> I see a fox
> Trying to eat me.
>
> Fox, fox,
> What does everyone see?
> We see the meadows
> Where we like to be.
>
> *Written and illustrated by Darren, Mia, Remy & Daniela*

d) *One Pig, Two Pigs* by Robert and Marlene McCracken (Tiger Cub Books), is a simple, whimsical beginning book that can be used as a model for factual writing. The text is fully illustrated. It begins:

> One pig on the table,
> Two pigs under the table.
> One pig in the car,
> Two pigs under the car.

Children can rewrite this factually with such statements as:

> One bear in the cave,
> Two bears by the cave.
> One bird over the nest,
> Two birds in the nest.

3. Write and Make Books

a) Make quick reading books.

i. Take dictation by asking about an animal that the children know well. Ask, 'What does a rabbit have?' Write each answer on an individual sentence strip. The answers for a rabbit might be:

four legs	long whiskers
a short tail	a twitchy nose
two long ears	soft fur

Place the sentence strips on top of one another. Make a cover sentence strip that says, **A rabbit has.** Staple the pages together and the child has a beginning reading book in which he reads the cover and then page one, reads the cover again and then page two, and so on.

ii. Similarly, books can be built about other rabbit attributes by asking the questions:

        What does a rabbit eat?
        What can a rabbit do?
        Where can a rabbit run?

Other animals can be substituted for a rabbit, and dozens of beginning books can be made. Children may wish to make their own individual books following the examples of the class books.

iii. Teach about one animal by reading factual books orally such as Sharks by Russel Freedman, *Puffin* by Deborah King, *The Fox, The Squirrel, The Beaver, The Spider, The Frog and The Fish* by Margaret Lane, *Biography of a Mountain Gorilla* by Lorele K. Harris, or *Gorilla, Rajpur: Last of the Bengal Tigers,* and others by Robert McClung.

iv. Brainstorm and list forty or more ideas on the chalkboard. Orally develop sentence sense by having children join two or three of the facts into a single sentence. If they can do this fairly easily, challenge them to create orally a short story about the animal. Again, if they can do the oral work easily, have them write short stories and illustrate them.

b) Make puzzle books

Children love to make puzzle books with four to five pages. They print a clue on each page and then illustrate the answer on the last page. The goal is to make the book so exactly right that everyone knows the answer. To do this with only three to four clues is a challenge. For example:

page 1. It is black and white.
page 2. It has white stripes.
page 3. It lives in Africa.
page 4. It looks like a horse.
page 5. (picture of a zebra)

or

page 1. It can sting.
page 2. It goes 'Zzzzz, zzzz, zzzz.'
page 3. It is dark brown or black.
page 4. It hatches from eggs laid in water.
page 5. (picture of a mosquito)

c) Make build-up-books for each animal habitat. Build-up-books can be made by the teacher to provide reading activities for the class or they can be made by children as a writing activity. They can be built as a flannel board activity for oral language development.

Begin by brainstorming all the animals that might live in a particular habitat. Now introduce the format using flannel board, pocket chart, or the chalkboard with simple illustrations. Draw a circle on the chalkboard and tell the children it is a pond. Ask what might be living in the pond and develop the answer, 'In the pond there is a duck.' Take a second response and develop the answer, 'In the pond there is a duck and a frog.' Now add several more inhabitants. Show the children how to create this book as illustrated.

In the Pond   In the pond there is a duck. (p. 1)   In the pond there is a duck and a frog. (p. 2)

Each page adds one thing:
Page 3: In the pond there is a duck, a frog and a turtle.
Page 4: In the pond there is a duck, a frog, a turtle and a fish.
Page 5: In the pond there is a duck, a frog, a turtle, a fish and a snake.

d) Create zero books. These are books for first grade children who are learning the concept of zero. Brainstorm for dangerous animals and where they live. Use these animals to write in the following pattern:

> There are 412 tigers in the jungle, BUT there are 0 tigers under my bed.
>
> There are 56 crocodiles in the river BUT there are 0 crocodiles in my school room.
>
> Children illustrate the zero page.

4. Write Lists

a) Brainstorm to get habitat lists and write using the lists. For example, ask "What animal might you hear . . ."

> on the way to school?
> in the forest?
> in the barnyard?
> at the seashore?

The answers could be written in many ways:

> On the way to school I heard . . .
> > a puppy barking,
> > a bird chirping,
> > a cat meowing,
> > a dog growling,
> > and a crow cawing.

Or: I heard _____
_____
> on my way to school.

> The teacher may ask "What animals are part of the . . ."
> > world under your feet?
> > world high above you?
> > ocean world?
> > world of a coral reef?
> > dark world of a cave?

Children can write in the pattern:

> A soaring eagle,
> A diving seagull,
> A hunting owl,
> A swooping hawk
> Are all part of the world high above me.

b) Work with an opposite idea. Ask the children to name places that are not good for some animals. List such places as trees, the jungle, the Arctic snow, burrows, nests in trees, nests in the grass. Write lists such as the following:

Trees aren't a good place for
    crocodiles,
    whales,
    hippopotami,
    turtles,
    fish,
    or sheep.

The jungle isn't a good place for
    puffins,
    penguins
    narwhals,
    walruses,
    polar bears,
    or caribou.

5. Sophisticate Understandings by Additional Sortings

a) Sophisticate children's understanding of habitat. Begin with stream, pond, river, lake, ocean, mountain, valley, and move on with veldt, desert, slough, prairie, swamp, tundra and delta. Then perhaps move on to specific areas such as Swiss Alps, Himalayas, Andes, Rocky Mountains, the Mississippi River, the Nile, the Amazon, or the Zambesi. Teach these terms and have the children use them in answering puzzles for the day as follows:

    Where in the world would you be if . . .
        penguins slid by?
        buffalo roamed near you?
        an emu raced past you?
        a tiger hid in the shadows?
        a koala bear climbed a tree?
        a kangaroo jumped over your head?
        a python slithered up a tree?
        a cockatoo flew overhead?
        a water buffalo came for a drink?
        a beaver slapped a warning with his tail?
        an anteater sniffed at a log?
        an armadillo waddled past?
        a parrot chattered in the trees?
        a road runner zoomed by?
        a capybara nibbled leaves?
        you saw a panda in a tree?
        a caribou clattered by?
        a wapiti bellowed?
        a polar bear slept on an ice floe?
        a llama carried a heavy load?
        a yak pulled a cart?
        a loon echoed its haunting call?

b) The puzzle for the day is placed on the classroom wall or door with a pocket underneath to hold answers that the children write on cards.

c) The child's task during the day is to search to find the answer, write the answer on a card, and place it in the answer pocket. At the end of the day the answers are read aloud and discussed. It would be good to have a world map or large globe available at this time to locate the various places that might be given as answers. The type of habitat should be discussed as well as the various countries in which this animal is found.

6. Write Stories
    a) Select one animal. Put the following six or seven headings on the chalkboard: **looks like, can do, has, lives, eats, enemies, young.** Brainstorm answers for each column. **What does it look like? What can it do? What does it have?** And so on. Try to get at least four items under each heading. Now create oral stories by having children use at least one item from each list to tell about the animal. For example with a beaver:

BEAVER

looks like	can do	have
rat-like rodent	swim very well	webbed back feet
thick brownish fur	hold breath for	scaly tail
sleek underwater	five minutes	long flat tail
weighs 40-50 pounds	gnaw down trees	big front teeth
30 inch body	carry things in	back toes for
	their front paws	combing
	work at night	

lives	eats	enemies
holes in banks	bark	wolves
lodges in ponds	roots of water	man
dam streams	plants	bear
lodge half under-	twigs	foxes
water		lynx
use mud, sticks		
and stones.		
leave breathing		
holes in top		
entrances under-		
water		

b) We can create a simple story now by using just one item from each column. To help children if they have difficulty, the teacher selects one child and tells him to read as she points. She points to some of the words (see the bold face print in the sample that follows) while the child reads orally and she fills in the rest orally as needed.

    The **beaver looks like** a big **rat** with a **long flat tail**. It **swims very well** and usually **lives** in a **lodge half underwater**. It **eats bark**. Its main **enemy** is **man**.

C. CLASSIFICATION

1. **Three Basic Ways to Classify Words**

    There are three basic ways to classify words:
    - By their internal structure.
    - By the way they are used in speaking and writing.
    - By the concepts they represent.

    The emphasis should be upon concepts and learning, but if children are to learn about language they should also give attention to the first two which are concerned with the form of the language.

a) The first task is to create a noun word bank of animal names. Brainstorm for all the animals children know. As you brainstorm classify the animals into several categories. For example, they may be listed by habitat. Domestic animals might go in one column, those that live in the Arctic snow in another, those that live high in the mountains in a third, those that live in salt water in a fourth, those that live in the desert in another, those that live in the plains of Africa in another, those that live in forests in another, and so on. The classifying will be somewhat inexact since some animals fit into more than one category because they migrate or because one member of a species, such as the polar bear, lives in the Arctic, while another, the black bear, lives in the forest. List the animals in more than one category, and strive for exact names within the species. We do not tell the children our classifications; we provoke them to solve our categories by asking, 'What else will fit in the column?' Classifying while brainstorming will cause the children to think and results in more animal names than random listing. We use one final classification, Cloud Nine. We put any animal we can't spell or any animal which we are unable to classify into it and assign children to look them up after the brainstorming is done for the day.

The following word bank came from a grade one/two in thirty minutes of brainstorming:

elephant	horse	beaver	whale	bear
leopard	cow	frog	shark	wolf
lion	sheep	turtle	tuna	deer
antelope	dog	snake	seal	coyote
jackal	chicken	heron	sea lion	cougar
rhinoceros		water	barracuda	
cheetah		moccasin	bass	
		bass	turtle	
			puffin	

kangaroo	worm	robin
koala	ant	sparrow
cockatoo	slug	bluebird
wallaby	beetle	owl
	centipede	hawk
		crow

Place all the animal names on cards. Have two sets made.

One set is hung (masking tape sticky side out is fast and easy), and the other set is used for classification activities. Add animals by teaching. One source is 'Animals Around the House' found in the Myself theme on pages 14-15. (Also available as a pocket card set. See bibliography.)

b) Present the following chant taken from the "SHAKER MANIFESTO OF 1882":

>Alligator, beetle, porcupine, whale,
>Bobolink, panther, dragonfly, snail,
>Crocodile, monkey, buffalo, hare,
>Dromedary, leopard, mud turtle, bear.
>Elephant, badger, pelican, ox,
>Flying fish, reindeer, anaconda, fox,
>Guinea pig, dolphin, antelope, goose,
>Hummingbird, weasel, pickerel, moose,
>Ibex, rhinoceros, owl, kangaroo,
>Jackal, opossum, toad, cockatoo,
>Kingfisher, peacock, anteater, bat,
>Lizard, ichneumon, honeybee, rat,
>Mockingbird, camel, grasshopper, mouse,
>Nightingale, spider, cuttlefish, grouse,
>Ocelot, pheasant, wolverine, auk,
>Periwinkle, ermine, katydid, hawk,
>Quail, hippopotamus, armadillo, moth,
>Rattlesnake, lion, woodpecker, sloth,
>Salamander, goldfinch, angleworm, dog,
>Tiger, flamingo, scorpion, frog,
>Unicorn, ostrich, nautilus, mole,
>Viper, gorilla, basilisk, sole,
>Whip-poor-will, beaver, centipede, fawn,
>Xanthos, canary, polliwog, swan,
>Yellowhammer, eagle, hyena, lark,
>Zebra, chameleon, butterfly, shark.

Add these names to the list as children prove they know something about each animal. The chant can be sung to the tune of 'Little Brown Jug.' Alice and Martin Provensen have illustrated delightfully a picture version of this chant in *A Peaceable Kingdom*.

Note the form of the chant. Each line has four words and with the exception of two lines ending in kangaroo and cockatoo, the lines all end with a word of one syllable. Any four words chant easily if the last syllable is accented. This is done by always putting a monosyllabic word at the end. Any set of four lines can be finished by a three-word repetition. For example:

>   Robin, bobolink, turkey, quail,
>   Sparrow, gold finch, whimbrel, rail
>   Eagle, sparrow hawk, ptarmigan, stork,
>   Sand piper, mallard, woodpecker, lark,
>   Birds, birds, birds!

Children can use this pattern to record and write from many of the classification groups they make as they sort the word cards.

2. Classification Through Structure

Through attention to the structure of the individual words we have children drill many phonic representations.

Children can classify words into the number of syllables, and with all except the monosyllabic words they can classify each group according to the accented syllable.

Children can find words that end in the same way, those which rhyme, and those that are alliterative.

3. Classification Through Use

Traditional grammar teaches nouns, verbs, adjectives, and so on. We suggest presenting words in slightly different ways before using the formal terms of grammar. To do this we prepare word and phrase strips taken from animal stories. We put the words and phrases on cards so that they fit the following headings:

> What animal it is.
> What the animal did or is doing.
> Where it did it.
> Why it did it.
> When it did it.

We include 'how' in some verb phrases as complete predicates, and some adjectives and descriptive phrases with the nouns as complete subjects. We offer the following list, but suggest that a list taken from stories that have been read to the pupils is more effective:

because it was dinner time	sheep	in a hollow tree
waited patiently	peacocks	watched
in the frozen Arctic	a dove	the playful dolphin
under the roots	a timid rabbit	a killer whale
throughout the grasslands	to rest and sleep	ran awkwardly
in their holes	in the starlight	a baby kangaroo
under houses	down in the mud	a huge python
deep in the tall grass	under the noon sun	a lioness

on the desert sand	on their nests	wandered
in a shallow pool	over the waves	turtles
to hide from the owl	in between rocky ledges	the king of the jungle
on the edge of the jungle	a fierce wild boar	graceful giraffes
through the vines	those bears	to bask in the sun
in the crisp, cool air	the noisy squirrels	in shallow streams
by the mouth of the cave	the deadly cobras	in flowery pastures
in the river's rapids	a herd of elephants	grazed
high in the mountains	by the meadow	the koala
in the depths of the ocean	twittered	after daybreak
up a tree trunk	slithered	over the grass
high overhead	curled	over the forest floor
screeched	protected	on the crest of a wave
soared	played	while the sun set
until it was safe	by the river's edge	high in the tree tops
almost midnight	a flock of flamingos	to listen intently
pink salmon	to find some fish	on the ice
until nighttime	to feed its young	deep in the jungle
as the sun shone brightly	yelled loudly	a black leopard
very late at night	crept	beavers
at lunch time	froze	a perky robin
while the babies played	crawled cautiously	striped chipmunks
during an instant of fear	seagulls	wrens
at the break of dawn	to run from the fox	many monkeys
at supper time	at their lodges	deer
in the heat of the day	to warm himself	a noisy bluejay
at a watering hole	to get cool	a huge python
as the sun rose	and share some dinner	porcupines
deep in the night	to catch its prey	two mallard ducks
to clean himself	as the babies slept	rested
to learn to fly	the next day	pheasants
because it was frightened	fuzzy caterpillars	hunted
camels	the hungry wolf	floated
the striped tiger	polar bears	and had his lunch
worms	a red fox	swam
crocodiles	the tiny mouse	flew
zebras	rainbow trout	squirmed
hummingbirds	at daybreak	pranced
pounced	crouched	trumpeted
strutted	frolicked	scurried

stalked	wriggled	tunneled furiously
nibbled	drifted	howled
barked	gnawed	dug
hippos	raced	chattered
under the light of the moon	jumped	scampered
trotted	boxed	fluttered
hopped	scuttled	swung
roamed	bathed	leaped
climbed	huge	
fished	skunks	
perched	flitted	

4. Activities For Working With The Word/Phrase Cards

    a) Make cards that say WHAT IT IS, WHAT IT DID, WHERE, WHY, and WHEN. Have the children sort the words and phrases to the cards. If the children find this difficult, omit WHY and WHEN, removing those cards from the pile until children classify the remaining cards with ease.

    b) Form sentences using all classifications:

        Throughout the grassland graceful giraffes grazed under the noon sun to bask in the heat.

        A flock of flamingos waited by the river's edge to get cool and share some dinner.

        At the break of dawn, deer ran across the forest floor to drink at the watering hole.

    c) Transform sentences.

        Sentences should be transformed physically. The children create a sentence using at least one card from each classification, and then move the cards to create the same sense with different word/phrase sequences. For example:

        By the river's edge, a flock of flamingos waited to get cool and share some dinner.

        A flock of flamingos waited to get cool and share some dinner by the river's edge.

        To get cool and share some dinner, a flock of flamingos waited by the river's edge.

d) Brainstorm

Use the **What it did** cards (the verbs) to brainstorm all the animals that might have:

>nibbled,
>hunted,
>dug furiously,
>slithered,
>trumpeted,
>sprayed,
>perched.

Use the **Why** phrases to brainstorm all the ways animals:
>get cool,
>protect their babies,
>warm themselves,
>hide from the owl,
>clean themselves.

5. Classification Through Concepts

a) Have children classify words in the same way they classified pictures. Put children in groups of four and allow them to classify freely. Then add a rule. For children at the beginning stages of reading and writing, make these rules very simple. For example, put all the animals that have four legs in one pile. All the other animals go in another pile.

As children classify, the rule is placed in the pocket chart in the form of a list and contrast:

>A _____ has four legs.
>A _____ has four legs.
>A _____ has four legs.
>A _____ has four legs.
>A _____ has four legs.
>But a _____ doesn't have four legs.

b) When the small groups have completed their classifications, their results are placed in the pocket chart list and read by the entire class:

>A bear has four legs.
>An elephant has four legs.
>A zebra has four legs.
>A lion has four legs.
>A giraffe has four legs.
>But a flamingo doesn't have four legs.

c) After the classifications of all small groups have been read by the entire class, children work individually, choosing their favorite animals and writing their own list and contrast. Thus, reading and writing activities flow freely from the classification activities.

6. Simple Classifications For Children Just Beginning To Read And Write

    Animals with whiskers, and those without.
    Animals larger than you are, and those which are not.
    Animals that live on land.
    Animals that are smaller than you are.
    Animals covered with fur.
    Animals with a tail, and those without.
    Animals that fly.
    Animals with horns.
    Animals covered with feathers.
    Animals with no legs.
    Animals that you can hold in your hand.
    Animals that live in water.
    Animals that live on a farm.
    Animals that frighten you.
    Animals that make good pets.

Shirley Rainey's first grade class in Langley, B.C. worked on an animal theme in November. One child, Kelly, wrote as follows on four consecutive days. The quality and amount of writing is not unusual when children become excited about their studies.

A spider is smaller than I am.	A skunk lives on the land.
A fly is smaller than I am.	A pig lives on the land.
A bunny is smaller than I am.	A rat lives on the land.
A pig is smaller than I am.	A cow lives on the land.
A turtle is smaller than I am.	A dog lives on the land.
A butterfly is smaller than I am.	A bunny lives on the land.
A rat is smaller than I am.	An ant lives on the land.
A frog is smaller than I am.	A horse lives on the land.
An ant is smaller than I am.	A cat lives on the land.
A bee is smaller than I am.	
A cat is smaller than I am.	
A duck is smaller than I am.	
A fish is smaller than I am.	
And an ant is the smallest animal I know.	

A cat has fur.	A bear has tail.
A puppy has fur.	A zebra has a tail.
A fox has fur.	A cat has a tail.
A bear has fur.	A rat has tail.
A sheep has fur.	A fox has a tail.
But a pig has no fur.	But an ant does not have a tail.

7. Classification Activities For Older Children Who Are Reading And Writing

This should develop both the concepts and vocabulary necessary for the study of animals. Ask questions that require sorting by concept. Introduce vocabulary through the questions. The following questions might be asked:

What animals might be . . .

    Sitting up on their haunches?
    Storing plant seeds in their face pockets?
    Walking elegantly with quick, small steps?
    Soaring in slow, wide circles?
    Leaping gracefully over low bushes?
    Roaming in the woodlands?
    Picking up their babies with their teeth?
    Carrying their babies on their backs?
    Twitching their whiskers nervously?
    Hanging by their feet?
    Scampering across the grass?
    Popping into a hole?
    Waddling into a den?
    Wriggling up a flower stem?
    Sliding on the ice and snow?
    Stalking in the grasslands?
    Leaping into the pond?

The following writings are from Shirley Rainey's second grade children in Langley, B.C. after they had done many sortings with cards:

> Starfish, clams and alligators can swim.
> Butterflies, bees and robins can fly.
> Slugs, snakes and worms can crawl.
> Kangaroos, rabbits and grasshoppers can hop.
> Horses, deers and foxes can run.
>
> Mammals are animals. You know they are mammals because they are warm-blooded.
>
> Mammals have fur or hair.
> Mammals are born alive. Mammals get milk from their mother.
> Mammals look after their babies.
> Dogs, horses, cows and elephants are mammals.
>
> Birds are animals with feathers.
> All birds hatch from eggs.
> Birds can lay eggs, fly and sing.
> Birds have wings, two feet, a beak and two eyes.
> Birds eat worms, fish and grains.
> Birds build nests to lay their eggs in and to sleep.
> Birds are fast, fuzzy and adorable.

8. Work With Nighttime Animals

The moon is out and the stars are twinkling. The animals of the day are silent. If you listen closely, you can hear the animals of the night. Some move around on the ground. Others climb in the trees, while still others fly.

a) Have the children classify all the animals they know as to whether they are awake or sleeping at night. The following animals are active at night:

fireflies	lions	ground squirrels
lightning bugs	tigers	owls
most moths	whip-poor-wills	rabbits
June bugs	nighthawks	opossums
frogs	kiwi	porcupines
lizards	grasshopper mice	skunks
bats	kangaroo rats	wolves
mice	pocket mice	beavers
rats	some snakes	raccoons
cats	coyotes	night herons
wild cats	badgers	

Add to the list by studying other animals.

b) Classify the nocturnal animals by their eating habits. Some eat meat, some eat seeds and grasses. Which animals are looking for food and which are using the darkness for protection?

c) Assign short research activities to study night animals:
1. How the animals protect themselves.
2. The special attributes of night hunters that enable them to catch prey.
3. Why so many desert animals are nocturnal.
4. How the smaller animals protect themselves from the night hunters.
5. Where they spend the day hours.
6. Do they ever come out in the day.

d) Brainstorm to make a small idea bank about one or more of the nocturnal animals. For example, with **Frogs,** brainstorm for adjectives, verbs, and where they might be. Add 'ing' to the verbs and practice the ideas orally in the following frame:

Frogs here, frogs there,
Frogs, frogs, everywhere!
_____ frogs _____
_____ frogs _____
_____ frogs _____
_____ frogs _____
Frogs _____
Frogs _____
Frogs _____
Frogs _____
Frogs! Frogs! Frogs!

e) Create chants such as:

Green frogs hopping,
Fat frogs croaking,
Young frogs leaping,
Old frogs sleeping,
Frogs in the garden,
Frogs in the pond,
Frogs on the lily pad,
Frogs near the swamp.
Frogs! Frogs! Frogs!

f) Write comparisons:

- Compare our human nighttime activities to those of the nocturnal animals:

At night I stay inside and watch TV while at the pond the frogs are croaking and laying eggs.

At night we _____ , while in the forest the squirrels _____ and _____.

- Compare the night behavior of the sleeping animals with the feeders and predators:

The bobwhites are crouched together in a circle on the ground while the fox searches for food in the underbrush.

- Use a repeated line to write poetically, and add a comparison on the last couplet:

The moon sends silvery beams into the forest.
Owl listens for a tiny sound from the forest floor.

> The moon sends silvery beams into the forest.
> Raccoon washes her food at the river bank.
>
> The moon sends silvery beams into the forest.
> Mouse searches for seeds in the forest shadows.
>
> The moon sends silvery beams into the forest,
> casting fluttering shadows on crow sleeping on a branch.

g) Read to children to discover interesting behaviors of sleeping animals. Have them research to discover others. Create a list of short paragraphs on a bulletin board, such as:

- Quail crouch together in a circle with their heads facing out. This is for protection. If an enemy sneaks up on them they immediately fly away, each one in a different direction. In this way, most usually escape.

- Birds roost. But they don't fall off their roost as they sleep. When a bird settles down to sleep on a branch, it tightens a cord in its legs, pulling its toes together and clamping them around the branch. The bird cannot fall; it has to lift its body to unlock its toes before it can fly away when it wakes up.

- Sometimes you can't tell when animals are sleeping. Animals such as snakes and fish have no eyelids so they sleep with their eyes open. Sometimes, fortunately very rarely, a person has eyelids that do not function properly and they sleep with their eyes wide open.

- Many animals such as horses, giraffes, elephants and hippopotami usually sleep standing up. Their knees lock so that their knees remain ridged, similar to the locking of the birds' claws.

- Some gorillas and orangutans build beds in trees. They snap off leafy branches and lay them across other branches to form beds.

h) Birds make a variety of nests. Have children classify birds as to where they nest and how the nests are constructed. This will take lots of research. Procure three or four abandoned nests of different types and have groups of children see if they can take the nest apart carefully to discover what materials were used in building it, and how the nest was constructed. Study the shapes and sizes of the nests.

i) Use the song *A Fox Went Out on a Moonlight Night,* a Weston Woods film. Learn the song and sort out the fact from fancy in the story. It is based on the book by Peter Spier.

j) Make charts and graphs of animals that:
sleep standing up,
sleep lying down,
perch to sleep,
sleep in trees,
sleep underground,
sleep in groups or herds,
sleep in the same place all the time or for several nights.

k) *National Geographic* presented a TV film on night animals, which is one of the many marvelous films they have made on animal life. We commend all of them to you for use with children, as well as their *World* magazine for children (National Geographic Society, Educational Services, 17th and M Sts. N.W. Washington, D.C. 20036). The following facts are taken from the film on night animals:

- Many animals that live in total darkness in deep caves underground are descended from prehistoric animals.
- In the pitch black darkness of the earth, there are many creatures who have never seen light and have never been seen by man.
- There are many creatures living in the depths of the ocean, in total darkness, which man has yet to study.
- The eyes of a cat gather ten times the light of the human eye, thus enabling it to see well at night.
- Owls' eyes are 100 times more sensitive than man's.
- A great horned owl swoops upon its prey guided only by its hearing.
- The bat is the world's only flying mammal. One group of bats devours 3 to 4 million insects in a single night, so bats are man's friends. Bats can hear ultrasonic sounds and they emit tiny squeaks that cannot be heard by human ears. For millions of years, bats have navigated through this sonar system, although we still do not understand how bats communicate or navigate. Scientists are studying to try and discover this. Some bats eat fruit. Others prey on lizards,

birds, and even other bats. Some bats pollinate flowers. The sonar signals of a bat can distinguish between a fish and a floating leaf. We are just now beginning to learn how a bat obtains its food. The bat's seemingly erratic flight is directed by its sonar system. It catches insects in the membranes of its wings and tail as it flies. Most bats are helpful to man, but in South America some cattle die from rabies and blood loss after attacks by the vampire bat. Vampire bats hide in caves during the day and forage at night, attacking the cattle. No other species of bat will live with the vampire. It lives on blood.

- The hyena had been regarded as a scavenger, eating the dead animals killed by other animals. It has proven to be a night hunter, competing with the lion for food. The hyena needs to eat four pounds of meat a day. A Dutch scientist has tracked the hyena at night and has discovered the hyena will kill for food or to protect its young. Lions seldom hunt themselves, preferring to plunder hyena kills. Lions and hyenas are ancient enemies.

- Science is trying to do for human beings what time has done for animals at night. Blind people are being fitted with sonic glasses to help them walk, a human parallel to the bat.

9. Brainstorm For Seasonal Awareness

a) Brainstorm many, many animal activities. Classify these as to when they are done, i.e. winter, spring, summer, fall. Use the brainstormed ideas in the following frames. Work orally first, then assign the frames as writing activities.

- The bear on the grassy mountain top is getting ready for fall. It's time for him to:

    grow a warm coat.
    find a sheltered cave.
    fill his belly.
    get ready for long sleep.

- The chipmunk on the forest floor is getting ready for fall. It's time for her to:

    build a two-room home underground.
    collect nuts and seeds.
    make a bed of twigs and leaves.
    grow a warm coat.

- The _____ (who, where) _____ is getting ready for fall. It's time for _____ to _____.
  _____.
  _____.

b) Contrast Summer and Winter Activities
  Use the frame:
  In summer _____ but in winter _____.

  Help children sophisticate their answers. Give them models to follow:
  In summer the snake slithers through our garden; but in winter he sleeps in a deserted hole in the ground.
  In summer the bear roams the forest eating berries, catching fish, and growing fat; but in winter he sleeps in a cave away from the snow.

c) Use Descriptions of the Season
  Brainstorm for other ways to say spring, summer, fall, and winter. Use these phrases in writing activities:
  As the winds roared and shrieked _____.
  By the time the last leaf fell from the tree _____.
  When the daffodils popped their heads through the ground _____.
  As the first few flakes of snow fell _____.
  On a sunny Easter morning _____.
  While the hot sun blazed down _____.
  The ice on the river was just breaking up when _____.

10. Brainstorm Animal Vocabulary
  a) Match the names of the adult animals with the names of their babies:

cat	kitten	lion	cub
cow	calf	seal	pup
owl	owlet	swan	cygnet
dog	pup	goose	gosling
fox	kit	horse	foal
hen	chick	moose	calf
bear	cub	sheep	lamb
wolf	whelp	tiger	cub
deer	fawn	whale	calf
duck	duckling	eagle	eaglet
frog	tadpole	oyster	spat
goat	kid	elephant	calf
hare	leveret	kangaroo	joey

b) Use the following poetic structures in chanting and then in writing:

> FIRST THINGS FIRST
> by Leland B. Jacobs
> A comes first, then B and C,
> One comes first, then two and three,
>   First things first.
>
> Puppy first, and then the dog,
> Tadpole first, and then the frog,
>   First things first.
>
> First the seed, and then the tree,
> That's the way it had to be,
>   First things first.

Put the structure on the chalkboard:

> First things first.
> First a _____, then a _____.
> First things first.

c) Encourage children to write chants and illustrate them.

> First things first.
> First a duckling, then a duck.
> First things first.
>
> First things first.
> First a calf, then a cow.
> First things first.
>
> First things first,
> First a cub, then a bear.
> First things first.

d) Randall Jarrell writes a delightful poem describing the life of a bat as part of his book, *The Bat Poet*. Jarrell begins his poem:

> A bat is born
> Naked and blind and pale.
> His mother makes a pocket of her tail
> And catches him. He clings to her long fur
> By his thumbs and toes and teeth.

Children use at least the first two lines to describe animals:

> a _____ is born
> _____ and _____ and _____.
> A duckling is born
> soft and downy and yellow.

>   A calf is born
>   spindly and wobbly and beautiful.
>
> Children may add a third and fourth line beginning with
>
>   Its mother _____.

e) Children may also match the animal and the noise it makes. Use the book, *Gobble, Growl, Grunt* by Peter Spier to help the children with this.

f) Make Brainstorming Predictions.
   Using their knowledge of animals, have children brainstorm to predict several endings to the following:
>   The frogs and toads grew silent as _____.
>   The owl flew from his branch to _____.
>   The fish hid behind seaweed when _____.
>   The deer leapt over logs and fallen trees when _____.
>   When he spotted the tiny fish glistening in the water _____.
>   The ants and small grubs that usually hid under the log scurried to _____.
>   _____ as the lioness sprang forward to secure her prey.
>   The spider got ready to wrap up her victim when _____.
>   _____ as they heard the cries of the wolf pack.

g) Create Similes.
   Brainstorm many answers for each of the following. Encourage children to go beyond the obvious. Read *Quick as a Cricket* by Audrey Wood. Work with the ideas orally before writing. Elicit as many oral answers as possible for each simile:
>   I love to see _____ as black as _____.
>   (horses as black as night.
>   kittens as black as jet.
>   leopards as black as licorice.
>   bears as black as ebony.
>   puppies as black as charcoal.)

Other Similes to use are:

_____ as fierce as _____
_____ as white as _____
_____ as tiny as _____
_____ as fat as _____
_____ as huge as _____
_____ as graceful as _____
_____ as sleek as _____
_____ as powerful as _____
_____ as brave as _____

Older children like to compare themselves to the qualities certain animals possess. Children write similes using the following frame:

I shall go through life as a _____, _____, _____, and _____.

I go through life as an eagle, proud, strong, and brave.

I go through life as a deer, swift, graceful, and sure. etc.

h) Brainstorm for reasons why children have a favorite animal. Why do they like them? Have children use their reasons in either one of the following frames:

i) Tell three things you like about your favorite animal, then finish your list for example:

> Soft, soft fur that's warm and grey.
> A soft, soft purr that's good to hear.
> A little pink tongue to lick my face.
> That's what I like about my kitten!

ii) Use the poem, *Because* by Eve Merriam. The poems begins.

> Because they hold their heads so high.
> Because their necks stretch to the sky.
> Because. . .(several more because statements)

Allow children to use the pattern of the poem:

Because _____.
Because _____.
Because _____.
Because _____.
Because _____.
Because _____.
Because, Because, Because.
That's why
I like _____!

iii) Develop the meaning of common sayings. Brainstorm for common sayings and develop their meanings orally:

a bee in his bonnet	a frog in her throat
like a dog with a bone	smell a rat
wait till the cows come home	lead a dog's life
beat a dead horse	let the cat out of the bag
in a pig's eye	horse feathers
rain cats and dogs	out of the horse's mouth

Children enjoy making a class book with literal illustrations.

D. DEVELOPING VOCABULARY AND SENTENCE STRUCTURE

1. Cloze Procedure

We use the Cloze Procedure orally to develop both concepts and vocabulary within any theme. We put the following Cloze exercise in the pocket chart or on the chalkboard:

One _____(1)_____ night, deep in the quiet _____(2)_____ the round moon _____(3)_____ through the leaves of an old oak tree. A large brown _____(4)_____ sat on the top of that _____(5)_____ and listened _____(6)_____ while a tiny mouse _____(7)_____ underneath him.

We ask children to read the entire story silently. We pronounce any word needed if a child has difficulty. We first ask children to decide whether this is part of a true story. We establish that the story is true. We begin the brainstorming by asking children to think of a word that would fit in the first blank. We write all the words on the chalkboard without comment, encouraging many words when it is appropriate to do so. We continue with each blank space, until all are done. The chalkboard might look like this:

(1)	(2)	(3)	(4)
dark	forest	shone	racoon
moonlit	woods	peeked	dog
frosty	meadow	glistened	bear
warm	pasture	glowed	squirrel
cool	stream	gleamed	cat
rainy	pond	glittered	owl
cloudy	mountains	glanced	hawk
Friday		sparkled	snake
spooky		twinkled	robin
beautiful		slipped	eagle
snowy		danced	porcupine

starry
summer
clear
winter
June
lovely

stepped
tip-toed
slid
peeped
winked

(5)
tree
log
chimney top
oak tree
rock
stone
old oak

(6)
intently
quietly
hungrily
wisely
curiously
anxiously
carefully
fleetingly
instinctively
worriedly
ravenously

(7)
ran
slept
scurried
froze
scampered
played
shivered
trembled
was trapped
hunted
searched
frolicked
shook
gnawed

This brainstorming completes the first lesson.

The next day, we check to see whether the words sound right, and whether they make sense. We chant all the words in the first column within the context, 'One night'. One dark night, One moonlit night, One frosty night, and so on. Having decided that all the phrases sound right, we check whether the words make sense. The children may decide that the words 'rainy' and 'cloudy' do not make sense because the story describes a quiet, moonlit night. When children give a logical explanation for why a word doesn't fit, we erase it.

Each blank is checked by the class. When children have finished, we might ask them to write the next paragraph of the story. Does the mouse get away, or does he provide the owl or the hawk with a good dinner?

Here are four more cloze exercises:
It was a _____ morning in _____. The _____ trees _____ from the wind which whistled through their _____ . A tiny _____ stepped from his _____ in the ground. He took two _____ steps, then his feet went out from under him and he _____ flat on his face on the _____.

Overhead, the sun looked like a yellow _____ in a patch of blue. The sand ran down to the _____, and on its slope were hundreds of _____. 'Crash!' came a _____ wave, and all of the little _____ ran back, unable to find the _____ they were eating for dinner. Presently, the sound of a _____ frightened the little _____ so much that they _____.

A little _____ lived in a _____. Tall grasses hid his _____ from _____. In a hollowed-out nest under an old _____ he slept during the day. Late one _____ the little _____ pushed through the leaves and grasses covering his _____ and started across the _____. He was tired of eating the same old food and seeing the same old _____ . Suddenly he saw a strange dark shape and heard the frightful sound of _____.

Foxes are _____. They move by _____ and spend the day _____ in a _____. Like cats, foxes _____ pretty well in the dark, and their hearing is much _____ than ours. Their sense of _____ is so _____ that their noses, quivering at the slightest smell, tell foxes all they need to know.

A good way to begin a story that you want to read orally to children is to make a cloze exercise using the first 50-100 words of the story, leaving out some critical words, and then predicting orally what will come next.

2. Teach Children How To Answer Questions.

Place a question in the pocket chart with one word per card:

| What | is | an | elephant | part | of | ? |

The capitalized form of 'An' should be on the reverse side of the 'an' card; the question mark should have a period on the reverse side. The question is answered orally, then the children transform the word cards of the question to begin their answer in the next pocket of the chart:

| An | elephant | is | part | of | | . |

Children complete the answer in their own spelling and with a variety of answers.

    An elephant is part of Africa.
    An elephant is part of India.
    An elephant is part of the zoo.
    An elephant is part of a herd.
    An elephant is part of the grasslands.

The answering of simple questions can be expanded into the writing of a short paragraph. Children answer each question, combining some answers into compound sentences. Much oral practise is necessary before successful writing takes place. Here are two examples:

This is a

It is part of something.
What is it part of?
What is it walking on?
Where is it going?
How many other feet does it have?

For example:
    This is a claw.
    It is part of an owl.
    It is walking on the branch
        of a tree, listening
        for sounds of the night.
It has one other feathered claw.

This is an

It's part of something.
Where is it?
What is it looking for?
Why?

Create your own lists to fit the animals or animal groups that you wish to teach about.

3. Structures for Beginning Writers

The following structures are simple to write because they require little independent writing. Most of this is copying so that can be done while children are beginning to learn to spell and write. The children may draw animal pictures instead of writing their names if writing is a problem.

    _____s here.
    _____s there.
    _____s, _____s everywhere.

    One _____.
    Two _____s.
    1, 2, 3, _____s.

A _____ that's tall,
A _____ that's small,
and that's all.

    This _____, that _____.
    Any _____, anywhere.
    I like _____.

I'd love to have
_____s, _____s, _____s, _____s
_____s, _____s, _____s, and _____s
in my zoo.

    _____, _____ in the sea.
    _____, _____ swim away.
    _____, _____ come back to play.

_____, _____ in a tree.
_____, _____ fly away.
_____, _____ come back to play.

    _____, _____ in a cave.
    _____, _____ go away.
    _____, _____ come back to play.

_____, _____ on the land.
_____, _____ go away.
_____, _____ come back to play.

There's a _____ in the _____.
There's a _____ on the _____.
There's a _____ by the _____.
    One, two, three.

Examples: Sharks here,
          Sharks there,
          Sharks, sharks everywhere.

One elephant.
Two elephant.
1, 2, 3 elephants.

        A dog that's tall,
        A chicken that's small,
        And that's all.

This fox, that fox.
Any fox, anywhere.
I like foxes.

        I'd love to have
        elephants, tigers, gorillas, lions,
        giraffes, a hippopotamus, and snakes
        in my zoo.

Whale, whale in the sea.
Whale, whale swim away.
Whale, whale come back to play.

        Sparrow, sparrow in a tree.
        Sparrow, sparrow fly away.
        Sparrow, sparrow come back to play.

Lizard, lizard in a cave.
Lizard, lizard go away.
Lizard, lizard come back to play.

        Horse, horse on the land.
        Horse, horse go away.
        Horse, horse come back to play.

There's a duck in the pond.
There's a deer on the meadow.
There's a gorilla by the tree.
One, two, three.

Younger children like to compare the abilities of animals with their own abilities. My first grade children loved this frame:

A _____ can _____.
He _____ all day.
But do you know what a _____ can't do?
A _____ can't _____ as I can.

A duck can swim	A robin can fly.
He swims all day.	He flies all day.
But do you know	But do you know
What a duck can't do	What a robin can't do
A duck can't sing 'Yankee Doodle'	He can't eat at McDonald's
As I can.	As I can.

4. Use Simple Poetry Structures to Increase Vocabulary.

a) Use an animal that children know very well to teach the procedure. Place the name of the animal in the pocket chart. In the next pocket add an adjective and repeat the animal name. In the next pocket add another adjective and repeat the previous line. Continue adding until you have created a four or five line poem.

> Kittens
> My kittens
> My two kittens
> My two white kittens
> My two soft white kittens

> Frogs
> Tiny frogs
> Three tiny frogs
> Three tiny green frogs
> Three tiny slimy green frogs

b) Diamante is a simple structure using adjectives, verbs, and nouns. We use it as follows:

> one word-noun
> two words-adjectives describing the noun
> three words-participles
> four words-nouns related to the subject

three words-participles
two words-adjectives
one word-synonym for the first word

For example:

Cats
Nice, kind
Playing, walking, running
Tail, legs, paws, fur
eating, licking, drinking
mean, rough
Siamese

E. WRITING STORIES FROM FRAMES AND STORY STRUCTURES

Pass out blank paper to students and have them fold the paper lengthwise. Have six colors of tempera paint mixed. The children drop a blob of paint in the center of each half. They should use two colors. They refold the paper and press the paint together, mixing it and spreading it. When the paint is dry, they add an outline, noses, feelers, legs, and so on to create a creepy crawler.

The children then write a factual story of their animal, using the answers to at least five questions:

Where was it born?
What is its name?
Where does it live?
What does it like to eat?
How big is it?
How does it protect itself?
What noises does it make?
How does it move?
Would it make a good pet?

Shirley Rainey's first and second grade children in Langley, B.C. studied elephants. They enjoyed using the following frame to invent fantasies about elephants:

Once there was a _____ elephant who had always wished to _____. One day a _____ came and granted the wish.
_____
_____.

The children practiced the story orally many times, suggesting many different wishes and consequences before they wrote individual fantasies. Two of their stories were:

> Once there was a little elephant who had always wished to fly. Then one day an elf came and granted his wish. So he was happy. He flew to White Rock. He had some fish and chips and he went to the beach. He was very happy. (Chris Gauvin, grade one)
>
> Once there was a wrinkly African elephant who wished he could be in a parade. One day a beautiful fairy came and granted his wish. After the fairy left, the elephant went shopping and bought a fancy blanket to put on his back when he was in the parade. One day there was a parade and the elephant was in the parade. But after a little while the elephant sat down. He slept and slept. Then he awoke. He got up and started to walk again. When the parade was over the elephant went on some rides. But all the rides he went on he broke them all. The elephant was sad. He did not want to be in a parade again and he didn't.
>
> (Jennifer Kehler, grade two)

F. ORIGIN STORIES

Origin stories are part of most folk literature and are available in many books and films.

Rudyard Kipling's *Just So Stories* are well known and offer, in written form, what the story teller might say to explain how animals got their various characteristics.

Dick Roughsey has written and illustrated *The Rainbow Serpent,* an aboriginal Australian story that explains how the various animals and birds came to populate Australia. This is available on a Weston Woods film. (Weston Woods Studios, Weston, CT. 06883, and Weston Woods Studios, Willowdale, Ontario, M2H 2E1)

Tomie de Paolo's *The Legend of the Bluebonnet* explains the origin of the bluebonnet. It is marvelously illustrated.

Other books in this area are:
How the Sun was Brought Back to the Sky, by Mirra Ginsburg.
The Girl Who Loved Wild Horses, by Paul Goble. (This is an American Indian legend which won the Caldecott Award for its illustrations.)
People of the Dreamtime, by Alan Marshall.
The Loon's Necklace, by Elizabeth Cleaver. (Available from Weston Woods as a film strip.)

Step 1

Read origin stories orally to children. This should be done over a period of 2-4 weeks with the emphasis on the enjoyment of the stories, with perhaps some note-taking on the content.

Notes should be obtained by brainstorming after the story has been read orally, perhaps after the second reading. We would brainstorm by asking the children to tell what they remember about the story. We take the responses and transfer them to sentence strips or a sheet of butcher paper, chanting as we take the responses on the chalkboard. Butcher paper will work best for comparisons later.

We teach paragraph writing with the sentence strips, usually working sequentially with this type of writing, so that the first step is to sequence the cards. The second step is to try several combinations of cards within the sequence to develop strong sentences. This is done orally, with the teacher acting as scribe, putting the best sentences on the chalkboard, creating a paragraph. If this is the introduction of paragraph writing to children they may merely copy it into their origin story notebook. However, for many children, the story should be erased and they should write their own versions from the notes. Later, the paragraph should be said orally only as a group, and even later only the most important ideas extracted by the group and put into the pocket chart.

It is probable that the teacher will wish to dismiss the abler writers once the main ideas have been extracted, and work with a few children who need repeated help in learning how to create good sentences and a good paragraph.

Children in late grade two and up might develop an origin story notebook by creating synopses. To do this, display all the responses from one story on strips and ask the children to select the five most important. This is to get the most important information on view. Use the strips selected to create oral paragraphs and then to write individual synopses.

Step 2

We hang the notes from at least five, and preferably ten, origin stories where the children can see them. We tell them that these are all origin stories, and ask them to see if they can decide what makes them so. What is their recipe? This is going to be fairly sophisticated work because there is no single recipe for an origin story. They should extract that origin stories do what the name implies: they tell how something we observe within the natural world came about.

After that, the words 'usually' or 'often' will have to be used. Often there is something magical or supernatural. Often there is a change in form. Often the change which should have been temporary is permanent. Usually, the change is sudden and irreversible and passed on to succeeding generations. Usually, the stories take place in the distant past well beyond anyone's memory or recorded history. This listing may take more than one study session.

Develop a recipe chart for origin stories, and list first those characteristics that are most important.

Provide thirty or more origin stories for children to read independently, from simple to difficult in reading level. Provide additional films and film strips of origin tales. If local native residents are available, see if one of their story tellers is available to come to class and orally tell some of their tales.

As children read, hold discussion groups to check the tale against the recipe. Focus on one part of the recipe within the group. One day, each child must find the part of their story which tells of the change and read it orally. We will check to see if the change is sudden and irreversible. Another day we can read the same tale or different tales to check for the intervention of the supernatural. We can assign children to be ready on Friday with a tale that has the intervention of the supernatural.

As we do this reading, we should develop several charts recording good story beginnings, typical endings, interesting words, the kinds of changes, the way in which magic is performed, and so on. In effect, we are collecting examples that validate the recipe.

Step 3

This is the writing of origin tales in imitation of those studied. It may be total group writing, and probably should be for almost all age levels if it is one of the first attempts. It may be writing within groups of four or five with the grouping being heterogeneous. It may be individual writing, but this would rarely be the only way or the first way.

The class and teacher should brainstorm for phenomena that might need explanation. For example:

> Why the dog has a bark.
> Why the octopus has eight arms.
> Why the camel has a hump.
> Why the kangaroo jumps.
> Why the zebra has stripes.

We might form writing clubs now by having children choose which phenomenon they wish to write about. We might choose one as a total class project so that we model the remainder of the process.

We follow our recipe and get a beginning. If we are going to explain why the zebra has stripes, then we need to begin by stating that once, long ago, the zebra was white. We need to state some of the problems of differences that this caused in nature. We need to set about some sequence of events that made the change, and we need an ending which states that ever since that time, zebras have had stripes.

Step 4

We may decide that we like our group story well enough to edit it properly, set it to pages, and assign groups of children to write the text carefully on a sheet of paper and illustrate the text to create a class book to be shared with other classes.

Following the group story, there may be additional brainstorming sessions for the purpose of helping those children who need help to generate ideas and to take notes on the chalkboard, which the child copies prior to setting to work individually.

There should be a sharing of books produced, and there should probably be creative drama with a narrator so that children can put their stories into dramatic form.

Sharon Lipscomb reads regularly to her grade one class in Custer, Washington, from Rudyard Kipling's *Just So Stories.* In January they wrote two Just-So-Stories. They did a marvelous job of imitating Kipling's style, his use of alliteration, his addressing the audience, 'Oh, Best Beloved,' his use of repetition, and even his use of inverted sentence syntax.

HOW THE SKUNK GOT HIS SMELL
Written by First Grade
(We made it up)
Illustrated by First grade
(We made the pictures)

Long ago, deep in the woods, Oh Best Beloved, the skunk had no smell.

There was a skunk named Silly, Sassy, Skippy Skunk. He had a bad habit. He ate everything he smelled. He would eat and he would eat and he would eat, eat, eat, EAT! Everytime he ate, he got spanked. He got spanked and he got spanked and he spanked, spanked, spanked, SPANKED! Everytime he got spanked he got mad. He got mad and he got mad and he got mad, mad, mad, MAD!

One day Silly, Sassy, Skippy Skunk smelled a spray can. He ate it whole. Then he got spanked by his father, spanked by his mother, spanked by his aunts, uncles, sisters and brother. He was mad.

Then he smelled something in Mr. McGregor's garden. So he went under the fence and ate a whole field of onions and garlic!

Now the more he ate, Oh Best Beloved, the more stuck the spray can got. Then most of what he ate went into the spray can. The can was full of skunk cabbage, onions, and garlic!

Now it happened whenever he was spanked, it set off the spray can. So when he was spanked by his father, spanked by his mother, spanked by his aunts, uncles, sisters and brother, they all got sprayed. Then THEY got mad and they got mad, mad, mad, MAD! So they never spanked him again.

"Where did you get that smell?" they all asked. Then all the skunks went out to get a smell. From that day to this, Oh Best Beloved, all the skunks have a smell that they can use when they get mad or spanked.

G. USING POETRY WITHIN AN ANIMAL THEME
1. The Nature of Poetry

The main characteristic of poetry is that it is written in lines. It has rhythm and sometimes rhyme, but poetry is distinguished by how the lines are put together to relate to one another. Writing lists is the simplest way to put lines together.

a) **List Poetry**
   i. The first line tells what the poem is about. The remaining lines list things about the topic. Lists are easy to write once the children understand the basic format. Supply the class with a list poem as a model. The following poems were modelled from 'Tree Climbing' by Kathleen Fraser.

   TREE CLIMBING
   This is my tree
   my place to be alone in,
   my branches for climbing
   my green leaves for hiding in
   my sunshine for reading
   my tree, my beautiful tree.

   This is my tree,
   My branches for running on,
   My leaves for hiding in,
   My hole for nesting in,
   My trunk for climbing.
   My tree, my beautiful tree!

   This is my seashore
   My surf to ride on,
   My tide pools to feed in,
   My shore for running on,
   My waves to wade in.
   My shore, my beautiful shore!

   The children read the poem, or two or three if they are the same model. Together they discuss and note how the lines are put together and how they relate to each other. The children are put in groups of four or five. They choose an animal habitat or are assigned a first line, and write a poem. Suggested beginnings are:

   This is my cave . . .
   This is my coral reef . . .
   This is my veldt . . .
   This is my prairie . . .
   This is my ice-floe . . .
   This is my burrow . . .

   ii. Children share ideas and write poems about particular animal habitats, such as the veldt, jungle, and so on. Each group is responsible for:
   a) creating a large mural background.
   b) naming animals living in that particular place.
   c) adding favorite animal figures to that background.
   d) writing a list poem for each favorite animal.

   Other list poems can be developed in a similar way by using the following ideas:

Brainstorm for what happiness is from an animal's point of view.

Happiness is	Happiness
My master calling me,	An old hay loft,
A clean basket to sleep in,	A field where rodents feed,
A bone for my supper,	The squeak of a mouse,
And a ball to play with.	A tiny animal in my claws,
	And a good, full tummy.

Use the form 'Happiness was' with Wilbur from *Charlotte's Web* as the author.

For example:

> "Happiness was"
> A good bucket of slops.
> Fern walking down the road.
> Charlotte to talk to.
> The quiet sounds of the barn.

Write of other animal characters from stories, both real and fantasy.

iii. List all the glorious things in an animal's life. Write a list using one or more of the following phrases listing at least three things before repeating the 'glorious' phrase:

> Glorious it is . . .
> Glorious it is to see . . .
> Glorious it is to hear . . .

For example:

> Glorious it is to see
> the snow slowly disappear from the forest floor,
> the new green shoots popping from the ground,
> the leaves budding.
> Glorious it is
> to nibble the new grasses,
> to feel the warmth of the spring sun on my back.
> to hear the robins' songs.
> Glorious it is.

iv. Lists can be used to record factual information. Read to children from animal books using language that children can understand. Aileen Fisher's *Anybody Home?* would yield a list of animal homes. The *Kids' Cat Book* tells thousands of facts about cats that would yield hundreds of lists. Ruth Heller's *Animals Born Alive and Well* would yield several lists of mammals and mammal characteristics. Children listen intently to the readings so that they can write factual lists such as:

> Elephants are the biggest land animals.
> Elephants have huge bodies.
> Elephants have little tails with little hairs at the end of it.
> Elephants have long trunks.
> Elephants have wrinkly skin.
> Elephants have big toenails.
> Elephants have tusks that are made of ivory.
> Elephants have big ears.
> Elephants have round legs.
> Elephants have little eyes.
> (Jennifer Kehler, Langley, BC Grade 2)

b) Lists Can be Sophisticated

i. Use a series to hold the list together. A series can be numerical as in a countdown, or it can be the days, the months, or the seasons of the year.

A countdown:

> At the bird sanctuary I saw
> One mallard duck flying by,
> Two white seagulls soaring high,
> Three pheasants nestling in the grass,
> Four Canada geese honking to pass,
> Five trumpeter swans out for a swim,
> And six blue herons looking slim.

The months of the year as a series can make the children think of an animal they might see in a particular locality each month of the year:

> In snowy January I spotted one grey squirrel.
> In frosty February I spied two frightened ground hogs.
> In windy March I saw three robins.
> In rainy April I spotted four fluffy rabbits.
> In warm May I spied five woolly lambs.
> In sunny June I saw six young calves.
> In the heat of July I spotted seven ducklings.
> In golden August I spied eight black bears.
> In cool September I saw nine striped chipmunks.
> In foggy October I noted ten Canada geese.
> In cold November I spied eleven elk.
> In crisp December I saw twelve white-tailed deer.

ii. Add adjectives, verbs, and prepositional phrases to your list poems.

> In the Arctic I saw:
> > A seal's nose popping through a tiny hole in the ice.
> > A huge polar bear wandering over snowy mounds.
> > A curious narwhal slipping through the icy waters.
> > A huge, heavy walrus diving from an icy ledge.

iii. Use a poetry model that requires both facts and description. Karla Kuskin's poem, "If I Were . . ." from *The Rose on My Cake* provides a delightful model. The following are three poems from twenty or more that were written by grade three children in Sharon Fortenberry's class in Burlington, Washington:

> IF I WERE A BEE
> I would buzz around
> And pollinate flowers.
> I would fly to my hive
> And make sweet honey.
> I would swarm with other bees.
> Being a bee. (Guy)

> IF I WERE A PANTHER
> I would run slyly
> And quietly climb trees.
> I would hunt spotted deer
> And hungrily eat dinner.
> Then sleep soundly in a tree.
> Being a panther. (Darron Drake)

> IF I WERE AN EAGLE
> I would fly and listen,
> Turning, looking, watching for prey.
> Listening for a twig snapping
> Or a leaf falling to the ground,
> Or a paw stepping in brown crunchy leaves.
> Being an eagle. (child's name not known)

2. Teach Metaphor

Metaphor is difficult for many children. We can make this understanding easier by teaching about metaphor through animals. Brainstorm for the names of cars and models listing them randomly. Have the children sort them, and sort out the animal names. (colt, pony, bronco, rabbit, jaguar, beetle, and so on)

Take two or more of the animals and list their attributes:

pinto:	rabbit:
colorful	quick
cute	fast
friendly	turns quickly
youthful	furry
hairy	soft
hungry	small
strong	alert
swift	lives in a hole

Ask the children which attributes the car maker hopes will come to mind when you are told to buy a pinto, a rabbit, and so on.

Have the children create a car, then a name that will best describe the car's characteristics. What kind of a car would be named elephant? gazelle? polar bear? killer whale? Choose the name for a car that is built to run economically, to run forever without needing repair, to start quickly, to run fast for a long time, to pull heavy loads, etc.

3. **Poetry Models For Writing Parody**
   - Margaret Wise Brown's poetry is probably the easiest to parody. "I like Bugs" is her best known, and she has used the pattern to write dozens of other poems.

      I like bugs.
      Black bugs,
      Green bugs,
      Bad bugs,
      Mean bugs,
      Any kind of bug,
      I like bugs.

      A bug on the sidewalk,
      A bug in the grass,
      A bug in a rug,
      A bug in a glass.
      I like bugs.

      Round bugs,
      Shiny bugs,
      Fat bugs,
      Buggy bugs,
      Big bugs,
      Lady bugs,
      I like bugs.

The following examples are from grade one pupils in Latimer Road School in Surrey, BC:

BUTTERFLIES
Colorful butterflies
Beautiful butterflies
Full grown butterflies
Squashed butterflies
Any kind of butterfly
I like butterflies.

A butterfly in a house
A butterfly in a plum tree
A butterfly in a hollow tree
A butterfly in a tree
I like butterflies.

Hairy butterflies
Lazy butterflies
Flat butterflies
Glass butterflies
Busy butterflies
Adult butterflies
I like butterflies. (Tyree M.)

BIRDS
Blue birds
Yellow birds
White birds
Black birds
Any kind of bird
I like birds.

A bird in a tree
A bird in a birdhouse
A bird in a cage
A bird in a house
I like birds.

Cute birds
Ugly Birds
Poor birds
Hungry birds
Born birds
Sleeping birds
I like birds. (Charlene G.)

BEAVERS
Busy beavers
Working beavers
Chopping beavers
Hungry beavers
Any kind of beaver
I like beavers.

A beaver in the woods
A beaver in the water
A beaver near the sea
A beaver on the ground
I like beavers.

Soft beavers
Eating beavers
Nice beavers
Fat beavers
Lazy beavers
Sleeping beavers
I like beavers. (Chris C.)

I LIKE FISH
Orange fish
Red Fish
Pink fish
Purple fish
Any kind of fish
I like fish.

A fish in the water
A fish in the ocean
A fish in the sea
A fish around me
I like fish.

Funny fish
Mommy fish
Baby fish
Daddy fish
Big fish
Gold fish
I like fish. (Angie)

- "What Do Sparrows Dream About?" by Robert A. McCracken can serve as a model.

  > Soft winds ablowing,
  > Pussy willows flowing,
  > Daffodils agrowing,
  > Pushing through the ground.
  > Heads tucked down,
  > What do sparrows dream about?

  The first four lines describe spring happening. Children can substitute other spring happenings or change the season entirely. The last two lines tell what an animal might be doing. Children choose their favorite animal and write a verse or verses for each season:

  > Winter winds roaring,
  > Bare trees shivering,
  > Storm clouds gathering,
  > Snow flakes falling,
  > In a hole under the snow
  > The red fox snores.

  > Hot sun scorching,
  > Dry grass withering,
  > Warm winds blowing,
  > Dandelions drooping,
  > In a hole deep in the ground
  > The red fox waits for the cool night.

- "Good Morning" by Muriel Sipe has a simple pattern for young children to use:

  > GOOD MORNING
  > One day I saw a downy duck,
  > With feathers on his back.
  > I said, 'Good morning, downy duck.'
  > The duck said, 'Quack, quack, quack.'
  >
  > One day I saw a timid mouse,
  > He was so shy and meek;
  > I said "Good morning, timid mouse,"
  > And he said, "Squeak, squeak, squeak."
  >
  > One day I saw a curly dog,
  > I met him with a bow;
  > I said, "Good morning, curly dog."
  > And he said, "Bow-wow-wow."

> One day I saw a scarlet bird,
> He woke me from my sleep;
> I said, "Good morning, scarlet bird,"
> And he said, "Cheep, cheep, cheep."

Line one tells what animal the author saw.
Line two tells something about the animal.
Line three tells what the author said to the animal.
Line four tells the animal's reply.

> One day I saw a little cat,
> With silky, soft, black fur.
> I said, 'Good morning, little cat.'
> The cat said, 'Purr, purr, purr.'

This pattern has a simple rhyme scheme which can be ignored. If you wish children to attempt the rhyme it would be best to brainstorm for what animals say, and then for words that rhyme with their sounds so that children can concentrate on making a sensible statement while using rhyme.

- Mary Ann Hoberman's "Fish" is a list about fish. It begins:

> Look at them flit
> Lickety split
> Wiggling
> Swinging
> Swerving
> Curving . . .

The poem continues adding participles. Children can parody this without a rhyming beginning, listing as many participles as are appropriate for the particular animal. They no doubt will find some rhyming pairs among their lists. Some rhyme is possible. Birds could begin: 'Look at them fly/Ever so high'; monkeys could begin: 'Look at them swing/From ring to ring': many animals could begin: 'Look at them race/Along the trace'.

- *Do Baby Bears?* is a short poetic book written and illustrated by Ethel and Leonard Kessler. It is a list with an ending contract. It contains whimsy and nonsense that children love. It begins:

> Do baby bears
> sit in chairs,
> comb their hair,
> wear underwear?
> No! But they do roll down the hill
> just as I do.

Children can write without rhyming such parodies as:

Do little puppies
wear big galoshes,
floss their teeth,
sleep in bunk beds?
No! But they do play with a ball
just as I do.

- David McCord's "Notice" provides a model that children love to parody. The first three lines each name an animal that the author has, and the last line tells where he keeps them. Grade one children in Sue Griffin's class in Burlington, Washington, wrote the following verses:

I have an ox,	I have a rat,	I have a dog,
I have a duck,	I have a bear,	I have a fox,
I have a fox,	I have a bat,	I have a frog,
Inside my truck.	Inside my hair.	Inside my socks.

I have a cat,	I have a cat,
I have an ape,	I have a goat,
I have a bat,	I have a bat,
Inside my cape.	Inside my boat.

- Rose Fyleman's "Mice" is in many anthologies. We put the poem in the pocket chart and have the children chant it many times. Then we brainstorm for all the attributes of a favorite animal and use the pattern as a list to parody:

   I think _____ are _____.
   Their _____ are _____.
   Their _____ are _____.
   Their _____ are _____.
   They _____.
   They _____.
   They _____.
   And no one seems to like them much.
   But I think _____ are _____.

- Vachel Lindsay's "There Was A Little Turtle" provides a simple model for writing about an animal. The first verse is the easiest to parody. Line one names the animal, line two tells where it lives, and lines three and four tell what it does. For example:

There was a little cub,		There was a little bird,
He lived in a den,		She lived in a nest,
He played in the mountains,		She dug up worms,
He fished in the streams.		She swallowed flies.

>There was a little monkey,
>He lived in the trees,
>He ate bananas,
>He swung from the branches.

- Margaret Wise Brown's "Old Snake Has Gone To Sleep" repeats the title as the second line of four couplets. Each first line describes a summer day.

>OLD SNAKE HAS GONE TO SLEEP
>Sun shining bright on the mountain rock
>Old snake has gone to sleep.
>Wild flowers blooming round the mountain rock
>Old snake has gone to sleep.
>
>Bees buzzing near the mountain rock
>Old snake has gone to sleep.
>Sun shining warm on the mountain rock
>Old snake has gone to sleep.

Children can write about their favorite animal and shift the season of the year to the season of their choice, or they can write two poems contrasting seasons:

>Winds roar loudly on the mountain top.
>Old bear snores in his cave.
>
>Black snow clouds swirl round the mountain top.
>Old bear snores in his cave.
>
>Snowflakes filter onto the mountain top.
>Old bear snores in his cave.
>
>Icicles form on the mountain top.
>Old bear snores in his cave.

- The traditional rhyme, "A Little Squirrel," yields a pattern for parody writing:

A LITTLE SQUIRREL
I saw a little squirrel
Sitting in a tree.
He was eating a nut
And wouldn't look at me.

This easily changes to:

I saw a little robin
Sitting in a tree.
He was eating a worm
And wouldn't look at me.

With a little more work it can be changed to:

I saw a _____
Swimming in the sea. And so on . . .

or

I saw a _____
Sitting on _____
He was eating _____
But he didn't see me. (With this last pattern ignore the lack of rhyme.)

- This countdown poem by James Reeves can be used as a writing pattern.

A PIG TALE
  Poor Jane Higgins,
    She had five piggins,
And one got drowned in the Irish Sea.
  Poor Jane Higgins,
    She had four piggins,
And one flew over a sycamore tree.
  Poor Jane Higgins,
    She had three piggins,
And one was taken away for pork.
  Poor Jane Higgins,
    She had two piggins,
And one was sent to the Bishop of Cork.
  Poor Jane Higgins,
    She had one piggin,
And that was struck by a shower of hail,
    So poor Jane Higgins,
    She had no piggins,
And that's the end of my little pig tale.

Set the pattern in the pocket chart as follows:

Poor _____ _____,
She had five _____,
And one _____.
Poor _____ _____,
She had four _____,
And one _____. And so on.

4. **Animal Poetry for Reading to Children:**
   - Suggested poems:

     "Mrs. Peck Pigeon" by Eleanor Farjeon
     "Under the Ground" by Rhoda Bacmeister
     "The House of the Mouse" by Lucy S. Mitchell
     "The Duck" by Ogden Nash
     "The Skunk" by Dorothy Baruch
     "Jump or Jiggle" by Evelyn Beyer
     "I Want You to Meet" by David McCord
     "Houses" by Ilo Orleans
     "Chant Number Four" by David McCord
     "Little Donkey Close Your Eyes" by Margaret Wise Brown
     "The Bat" by Frank Jacobs
     "The Crocodile" by Lewis Carroll

   - Here are two poems by anonymous writers. Children will enjoy their fun.

     I'VE GOT A DOG

     I've got a dog as thin as a rail,
     He's got fleas all over his tail;
     Every time his tail goes flop,
     The fleas on the bottom all hop to the top.

     IF YOU EVER MEET A WHALE

     If you ever, ever, ever, ever,
       ever meet a whale,
     You must never, never, never, never,
       grab him by his tail.
     If you ever, ever, ever, ever,
       grab him by his tail _____
     You will never, never, never, never,
       meet another whale.

   - The Tiger Cub Book, *Should You Ever?* by Robert and Marlene McCracken, is a beginning book that takes advantage of the previous pattern.

- Poetry and Chants About Bears

    THE BEAR SONG

    The bear went over the mountain,
    The bear went over the mountain,
    The bear went over the mountain,
    And what do you think he saw?

    He saw another mountain,
    He saw another mountain,
    He saw another mountain,
    And that was all he saw.

Using the Bear Song:

Place the phrases in the pocket chart. Chant and track with the children until they are familiar with the words. Then change the poem and rewrite:

Change the verb "went" to other verbs that would tell how a bear moves (stomped, lumbered, charged, raced, and so on.)

Change what the bear would see. What might be on that other mountain? (a great big sheep, a mountain goat, a herd of deer, a small red fox, and so on.)

Change the song to use various senses. The bear went over the mountain to see what he could taste or smell (some chili cooking, some bees' honey, and so on.)

- Make this into a big book to chant and dramatize:

    TEDDY BEAR, TEDDY BEAR

    Teddy Bear, Teddy Bear
       turn around,
    Teddy Bear, Teddy Bear
       touch the ground,
    Teddy Bear, Teddy Bear
       take a bow,
    Teddy Bear, Teddy Bear
       show us how,
    Teddy Bear, Teddy Bear
       row a boat,
    Teddy Bear, Teddy Bear
       button your coat,
    Teddy Bear, Teddy Bear
       give me a hug,
    Teddy Bear, Teddy Bear
       catch a bug,
    Teddy Bear, Teddy Bear
       read a book,
    Teddy Bear, Teddy Bear
       try to cook,
    Teddy Bear, Teddy Bear
       hop like a bunny,
    Teddy Bear, Teddy Bear
       act very funny,
    Teddy Bear, Teddy Bear
       bark like a dog,
    Teddy Bear, Teddy Bear
       croak like a frog,
    Teddy Bear, Teddy Bear
       go to bed,
    Teddy Bear, Teddy Bear
       cover your head,
    Teddy Bear, Teddy Bear
       turn out the light,
    Teddy Bear, Teddy Bear
       say, 'Goodnight!'

(Available as a big book and as a Tiger Cub Book) See bibliography.
- Going on a Bear Hunt

This is an echo song. The teacher leads by saying a line, then the children repeat the line. A rhythm is established before the chant begins by having everyone tap their feet on the floor in time to a given beat. This is contained throughout the chant.

LEADER	RESPONSE
Let's go on a bear hunt	Let's go on a bear hunt
There's a tree	There's a tree
Can't go over it	Can't go over it
Can't go under it	Can't go under it
Can't go around it	Can't go around it
Got to climb it	Got to climb it
	(motions of tree climbing)
There's a field	There's a field
Can't go over it	Can't go over it
Can't go under it	Can't go under it
Can't go around it	Can't go around it
Got to go through it	Got to go through it
	(Rub hands together to create "grass" effects.)
There's a river	There's a river
Can't go over it	Can't go over it
Can't go under it	Can't go under it
Can't go through it	Can't go through it
Have to swim it	Have to swim it
	(Make swimming motions with arms)
There's a thick woods	There's a thick woods
Can't go over it	Can't go over it
Can't go under it	Can't go under it
Can't go around it	Can't go around it
Have to go in it	Have to go in it
Oh-O-O-O-it's dark in here	Oh-O-O-O-it's dark in here
Oh-O-O-O-I feel something	Oh-O-O-O-I feel something
Oh-O-O-O-it's huge and furry	Oh-O-O-O-it's huge and furry
It's a BEAR!	It's a BEAR!

(Quickly run home, make feet tap much faster, arms move across the river; hands rub going through the field; make climbing up and down motions going over trees.)

H. BOOKS TO USE AS MODELS FOR CHILDREN'S OWN WRITING

- **"One Two Three Four," from *Sounds of Numbers* by Kate Consodine & Ruby Schuler.**

  The book begins:
  > In the first month of the year, I found one brown pony and he followed me home. In the second month of the year, I found two white kittens and they followed me home.

  Children in first and second grades can use this as a writing pattern. Combine the animal theme with the use of the calendar.

  In the _____ month of the year, I found _____ _____ and they _____.

- ***One Hunter* by Pat Hutchins.**

  This book is a countdown. It is told in pictures with very few words. One hunter passes one elephant, then two giraffes and so on. The idea of the countdown could be used for writing and illustrating in many ways.

- ***Rosie's Walk* by Pat Hutchins.**

  This popular beginning book is an exciting chase between a fox and a hen. The words are very simple, telling where the hen went. Almost all young children can use the pattern of this book to take their favorite animal on a walk. The following example is from John Perpich's first grade classroom in Vancouver, B.C.:
  > Leggy the octopus went for a swim,
  > across a river
  > past the seaweed
  > over a salmon
  > around the treasure
  > through a cave
  > under a whale
  > and got home before dinner.

- ***Good-night Owl* by Pat Hutchins.**

  This book could be used when studying nocturnal and diurnal animals. Owl is trying to sleep and all the animals of the daylight make noises and wake him up. Other animals could be substituted and worked into the pattern:
  > Hawk tried to sleep.
  > The dogs barked, bow-wow, bow-wow,
  > And hawk tried to sleep.
  > The cats meowed, meow, meow,
  > And hawk tried to sleep.

- *Going for a Walk* by **Beatrice Schenk de Regniers.**

    Here is a story that can be built and practiced very nicely in the pocket chart. The pattern repeats throughout the book. Once children have worked with the book in the pocket chart they find it very easy to write a version of their own. The book begins:

    > The little girl goes for a walk.
    > She sees a cow.
    > The little girl says HI!
    > The cow says MOO.
    > The little girl walks on.

The writing pattern is:

> She sees a _____.
> The little girl says HI!
> The _____ says _____.
> The little girl walks on.

- *I was Walking Down the Road* by **Sarah E. Barchas.**

    The book takes a rhyming walk. Children who are able to rhyme can use this pattern to write their own versions. The book begins:

    > I was walking down the road.
    > Then I saw a little toad.
    > I caught it.
    > I picked it up.
    > I put it in a cage.

The pattern merely works with the first two lines:

> I was _____.
> Then I _____.
> I caught it.
> I picked it up.
> I put it in a cage.

- *The Very Hungry Caterpillar* by **Eric Carle.**

    This book is a series. It uses the days of the week to describe an animal's eating habits. Brainstorm for favorite animals, then list at least seven things each of those animals eats. Children chant the lists in the following oral pattern:

    > On Monday he ate _____ but he was still hungry.
    > On Tuesday he ate _____ but he was still hungry.

After chanting, children will find it easy to use the pattern in writing.

- *At Mary Bloom's* **by Aliki.**

    Mary Bloom loves animals. She has a house full of them. Begin to use this book by brainstorming for all the animals that make good pets. Have children use all the animals they can think of not used in Aliki's version. Retell the story using the new animals:

    > The horse will neigh.
    > The canary will twitter.
    > The lizard will leap.

- *The Longest Journey In The World* **by William Barrett Morris.**

    An excellent story to use while children are learning about small animals or "bugs". We build it in the pocket chart in both words and pictures, allowing the entire class to enjoy the story several times. Then we choose other small animals from our word bank and tell the story again, using different animals and telling different places they went. We do this several times before we ask children to write their new story. Jennifer is in first grade at Kinvig School in Surrey, B.C. She wrote this story in April:

    > One morning as the sun was coming up, a little bunny said to himself, "I am going on a long journey." He hopped and he hopped and he hopped. He hopped over the bridge. He hopped on the swings. He hopped by the sand. He hopped near the teeter-totters. He hopped and he hopped and he hopped. That night as the sun went down, the little bunny wondered how far he had come. So he climbed a tall flag pole to look back. "I am truly amazed," he said to himself, "This is the longest journey in the world."

- *The Important Book* **by Margaret Wise Brown.**

    Children list (in phrase form) everything they know about their favorite animals. They use the structure of *The Important Book* to write these facts in a very simple paragraph form. Here are two second-grade examples from Mrs. Jarvos' classroom at Riverdale Elementary School in Surrey, B.C.

    Stuart wrote:

    > The important thing about a bald eagle is it has powerful wings. It is true that it catches mice and circles in the sky. It is true that it can see far and has sharp claws. But the important thing about a bald eagle is it has powerful wings.

    Kevin wrote:

    > The important thing about a kitten is it is huggable. It is true that a kitten is soft and cuddly. It is true that a kitten can play and jump. But the important thing about a kitten is it is huggable.

- *My Cat Likes To Hide In Boxes* **by Eve Sutton.**

    This book is popular with children. They write about their favorite pet, telling various antics that pet gets up to.

- *Monday I Was An Alligator* by **Susan Pearson.**
  Each day of the week, Emily pretends she is a different animal. This is a good book to re-write.

- *Borrowed Feathers and Other Fables* by **Freire Wright.**
  These animal fables provide a good pattern for children writing their own fables.

- *But Ostriches* by **Aileen Fisher.**
  This book is filled with patterns of lists, contrasts and rhyme. Many parts could be used as writing patterns.

I. CHANTS AND SONGS TO USE WITHIN AN ANIMAL THEME

- *Five Little Chickadees,* **a Traditional American Countdown.** This chant can be used as a fingerplay with each finger representing a chickadee:

                Five little chickadees peeping at the door,
                One flew away and then there were four.

Chorus   Chickadee, chickadee, happy and gay,
                Chickadee, chickadee, fly away.

Verse   Four little chickadees sitting on a tree,
                One flew away and then there were three.
                            Chorus

Verse   Three little chickadees looking at you,
                One flew away and then there were two.
                            Chorus

Verse   Two little chickadees sitting in the sun,
                One flew away and then there was one.
                            Chorus

Verse   One little chickadee left all alone,
                It flew away and then there were none.
                            Chorus.

- The following four poems by Robert McCracken can be easily memorized by young children. Put the text on word cards or phrase cards and build in the pocket chart to teach them as a way for children to practise with print.

THE BLUE WHALE

The big blue whale is huge and fat.
It swims about like this and that.
It has no fingers. It has no toes,
But with its flukes it really goes.

THE GIRAFFE

The spotted giraffe is tall and skinny.
It feeds on tree tops in the spinney.
It has no larynx to do its speaking,
But with its long legs it can go streaking.

THE KANGAROO

The kangaroo is big and fat.
She jumps about like this and that.
She's cautious with what's in her pocket,
Because she has no lock to lock it.

THE CROCODILE

The crocodile has lots of guile.
He's always pleased to meet you.
But don't be fooled by his toothy smile.
He really plans to eat you.

- ***Over in the Meadow,* by John Longstaff. A Traditional English Chant.** This excellent song should be built in the pocket chart and sung many times before it is used as a writing pattern:

    Over in the meadow in the sand, in the sun,
    Lived an old mother frog and her little froggie one.
    "Croak!" said the mother; "I croak," said the one.
    So they croaked and they croaked in the sand, in the sun.

    Over in the meadow in the stream so blue,
    Lived an old mother fish and her little fishies two.
    "Swim!" said the mother; "We swim," said the two.
    So they swam and they swam in the stream so blue.

    Over in the meadow in the branch of the tree,
    Lived an old mother bird and her little birdies three.
    "Sing!" said the mother; "We sing," said the three,
    And they sang and they sang on a branch of the tree.

- ***Where, Oh Where Has My Little Dog Gone?* An American Minstrel Song.**

    Oh, where, oh where has my little dog gone?
    Oh where, oh where can he be?
    With his ears cut short and his tail cut long,
    Oh where, oh where, can he be?

- *Go Tell Aunt Rhody,* **A traditional American song.**

    Go tell Aunt Rhody,
    Go tell Aunt Rhody,
    Go tell Aunt Rhody,
    Her old grey goose is dead.

    The one she's been saving,
    The one she's been saving,
    The one she's been saving,
    To make a feather bed.

    She died in the millpond,
    She died in the millpond,
    She died in the millpond,
    Standing on her head.

    The goslings are mourning
    The goslings are mourning,
    The goslings are mourning,
    Because their mamee's dead.

    The barnyard is weeping,
    The barnyard is weeping,
    The barnyard is weeping,
    Waiting to be fed.

    Now she's departed,
    Now she's departed,
    Now she's departed,
    Her old grey soul has fled.

- *Old Macdonald Had a Farm,* **Traditional English-American song.** This builds in the pocket chart either in words or pictures:

    Old Macdonald had a farm, E-I-E-I-O!
    And on this farm he had a dog, E-I-E-I-O!
    With a bow-wow here, and bow-wow there,
    Here a bow, there a bow, everywhere a bow-wow,
    Old Macdonald had a farm, E-I-E-I-O!

    And on this farm he had some ducks, E-I-E-I-O!
    With a quack, quack here, and a quack, quack there,
    Here a quack, there a quack, everywhere a quack quack.
    With a bow-wow here, and a bow-wow there,
    Here a bow, there a bow, everywhere a bow-wow.
    Old Macdonald had a farm, E-I-E-I-O!

The song is cumulative and can be continued until all of the animals in the barnyard have been named.

    chick — peep
    cow — moo
    pig — oink
    horse — neigh
    lamb — baa
    mouse — squeak
    donkey — hee-haw.

Children can practice matching animal pictures to the noises on word cards. (Available as a picture card set with some word cards. See bibliography)

- ***The Whale Song,* Traditional American Folk Song.**

    There are many regional variations of this. It is sung to the tune of "Dixie." Our version is:

    > In Puget Sound there lives a whale.
    > She eats pork chops by the pail,
    > by the pillbox,
    > by the bathtub,
    > by the pick-up,
    > by the seiner.
    >
    > Her name is Sarah, and she's a peach,
    > but you can't leave food within her reach,
    > nor nursemaids,
    > nor airedales,
    > nor chocolate ice-cream sodas.
    >
    > She eats a lot and when she smiles,
    > you can see her teeth for miles and miles,
    > and her tonsils, and her adenoids,
    > and things too fierce to mention.
    >
    > So what can you do with a whale like that?
    > What can you do if she sits on your hat,
    > or your toothbrush, or your grandmother,
    > or anything else that's helpless?

    (Available as a Big Book and as a Tiger Cub Book. See bibliography)

- ***Bill Grogan's Goat,* Traditional American.**

    This is an echo song. One person sings as leader, and the rest echo the lines:

Bill Grogan's goat	*(echo)*	*Bill Grogan's goat*
Was feeling fine		*Was feeling fine*
Ate three red shirts		*Ate three red shirts*
From off the line.		*From off the line.*

Bill took a stick . . .	He untied Bill's goat. . .
Gave him a whack . . .	From the railroad track. . .
And tied him to . . .	And took the goat. . .
The railroad track . . .	To Bill Grogan's shack. . .

The whistle blew . . .	He wore Bill's shirts. . .
The train drew nigh . . .	They fit just fine . . .
Bill Grogan's goat . . .	And chugged away. . .
Was doomed to die . . .	On the railroad line. . .

He gave three groans . . .
Of awful pain . . .
Coughed up the shirts . . .
And flagged the train . . .

Bill fed his goat. . .
with special care. . .
Lest the hungry goat. . .
Eat his underwear . . .

The engineer . . .
stopped the train in time . . .
He took the shirts . . .
Washed out the grime . . .

- Tongue Twisters (Available as a Big Book and as a Tiger Cub Book. See bibliography)

    How much wood would a woodchuck chuck
    If a woodchuck could chuck wood?
    A woodchuck would chuck all the wood he could chuck
    If a woodchuck could chuck wood.

    Swan swam over sea.
    Swim, swan, swim.
    Swan swam back again.
    Well swum, swan!

    Three grey geese in a green field grazing.
    Grey were the geese, and green was the grazing.

    A skunk sat on a stump.
    The skunk thought the stump stunk.
    And the stump thought the skunk stunk.

    A fly and a flea in a flue
    Were imprisoned, so what could they do?
        Said the fly, "let us flee!"
        "Let us fly," said the flea,
        And they flew through a flaw in the flue.

These are fun to write on word cards, mix up, and rebuild in the pocket chart.

ART ACTIVITIES ON AN ANIMAL THEME

Art Section by Wyn Davies

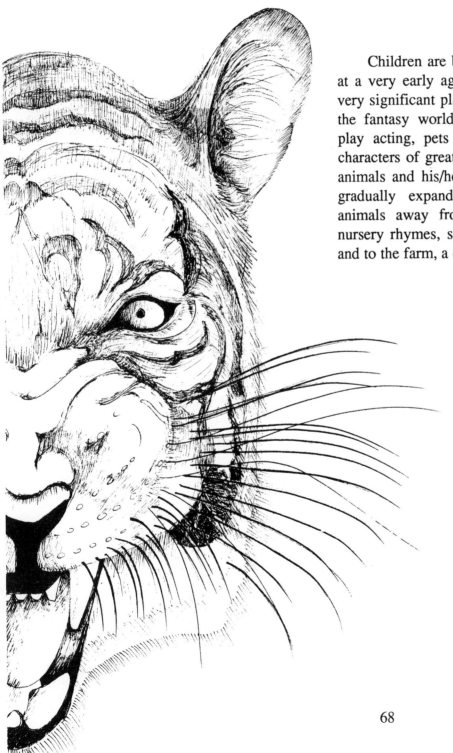

Children are brought into contact with animals at a very early age. A child's first pet occupies a very significant place in the child's world. Through the fantasy world of imaginary conversation and play acting, pets are frequently transformed into characters of great variety. A child's knowledge of animals and his/her familiarity with their habits is gradually expanded to include information on animals away from the home. Through poems, nursery rhymes, stories, film and visits to the zoo and to the farm, a child's knowledge and experience become enriched. Young children relate very closely to animals of all kinds and are keenly interested in their activities. The selection of an animals theme would guarantee a high degree of interest and enthusiasm.

In all the activities which follow, students should be encouraged to be imaginative and interpretive in both the areas of image development and in the handling of materials. The image of the animal should be expressed in a wide variety of ways as the theme develops, and the activities selected should be those which encourage thought and experimentation. The world of advertising is full of stylized, cartoon images of animals. The art programme should try hard to foster original thought.

The classroom should always contain charts and file cards which students assist in compiling together with a mini-library of materials on the theme being studied. Today, photographic images rather than drawn images exert the major influence on young children's images. Television, videos, film, cartoons, and so on, are all very influential components in developing the visual literacy of young minds. Therefore, it is important to provide as much 'face to face' contact with real situations–visits to farms, zoos, riding stables and pet shops. All these experiences would enrich the children's visual imagery beyond the superficial level of commercial art.

To supplement such field trips, begin charts and file cards early in the school year so that the shapes, surface patterns and the environments of various animals become increasingly familiar. Animal images might be classified this way:

- from the past
- fantasy animals
- farm animals
- circus animals
- household pets—real or make-believe
- animals in stories, poems and nursery rhymes
- animals at the zoo
- animals in motion pictures
- animals from different lands
- animals in works of art
- animals from historical cultures, and so on

Brainstorm ways of expanding this list with the class.

A. THE IMAGE EXPRESSED IN TWO DIMENSIONS

(1) Drawing and Painting

a) Arrange to have a child's pet visit the classroom as a subject for drawing exercises. Although the pet is allowed to wander around the room while the drawing takes place, students are still able to record characteristics of the pet far more sensitively than from memory. Draw in black felt pens on large sheets of paper. Exhibit the line drawings together with stories and snapshots of the pet brought from home.

Remember to give students an opportunity to draw the animals they see on field trips. Have a class set of small drawing boards made, sufficient to accommodate 9" x 12" paper for young children and 12" x 18" for older students.

b) As an extension of the line drawing exercise, cut out the drawings and mount them on a sheet of brightly coloured construction paper. Add to the background, details of the pet's real or imaginary environment. Some examples are:

- a rabbit in a rabbit hutch
- a rabbit in the wild—include flowers, insects, etc.
- a dog in a backyard—include plants and people
- a dog on a city street—include buildings, vehicles and people
- a cat in a living room—include furniture and people
- a cat in the jungle—include tropical plants, birds and insects

Use coloured felt pens, oil pastels or paint to complete the composition.

c) Draw a household pet from life or memory and later paint in bright tempera colour. Encourage the students to use bright imaginary colours such as a lime green cat, a purple dog, and so on. Cut out the image and mount on a sheet of brightly coloured construction paper. In the background, create a fantasy setting for the pet. Some examples are:

- my dog rides a rocket to the moon

- my dog goes sailing
- my rabbit rides the school bus
- my cat the traffic warden

Experiment with proportion so that the pet becomes miniature or gigantic. The scale of the background could also be life size or totally distorted.

d) Combine drawings of animals with any studies currently under way on various countries or cultures. For example:
- studies of the Arctic would naturally extend to include drawings and sculpture of seals, polar bears, etc.
- Australia—kangaroo, wallaby, koala bear, etc.
- Africa—lion, elephant, rhinoceros, etc.
- Canada—deer, bear, moose, elk, etc.

Use the drawings as banners throughout the hallways. For example, when Africa is being studied, glue big bold drawings or silhouettes of African animals to large sheets of construction paper. Staple the sheets back to back and hang from beams or ceilings. White silhouettes mounted on black construction paper is an effective way of doing this.

e) Draw a household pet from memory and include as many details as can be remembered of the pet's usual surroundings. Memory drawing is an essential component of the art programme since it trains students to be observant and perceptive in their visual recall.

Enrich the drawings later by adding colour in real or imaginary colour schemes. Give some direction on the choice and use of colour. For example:
- my dog in a world of green
- my cat in a purple garden
- my rabbit in a black and white back yard

Try some simple colour exercises first. With some white and green paint, how many shades of green can we mix? Draw the background for the animal in black felt pen and add the shades of colour later.

f) Examine photographs of animals in magazines and books. Develop large scale drawings on white paper and later cut them out and glue to a long piece of coloured mural paper. Add white cut-paper images for the setting and further embellish those images with black felt pens, emphasizing the patterns on the images. For example:
- the tiles on a roof
- the bark on a tree

- the leaves on a tree
- the boards in a fence
- the pebbles on a beach
- the waves in the sea
- the clouds in the sky

Develop repeating patterns throughout the cut-paper background and also on the animal.

g) After reading a story or nursery rhyme involving an animal, create a painting of the animal and outline it later in black felt pen. Complete the picture with coloured felt pen images to tell an incident in the story. Some possible sources might include:
- 'Little Red Riding Hood'
- 'The Three Bears'
- 'Dick Whittington'
- 'The Three Pigs'
- 'The Gingerbread Man'
- 'The Jungle Book' by Rudyard Kipling
- 'Sylvester and the Magic Pebble' by William Steig
- 'Bears' by Ruth Krauss
- 'Curious George' by H.A. Rey
- 'Blueberries for Sal' by Robert McCloskey
- 'Danny and the Dinosaur' by Syd Hoff
- 'The Hungry Crocodile' by Roald Dahl

Teachers will of course have their own favourites to add to this list.

h) Make drawings of imaginary pets and place them in a domestic setting. Make the drawings quite realistic using charts and books as research material. Then place the animal in an unusual situation. Some examples are:
- a dinosaur as an adventure playground
- an ant-eater as a rototiller
- a baby elephant as a vacuum cleaner
- a giraffe as a tree fort
- a crocodile as a giant nutcracker

Experiment with the scale of the animal and its setting.

i) Make a strong line drawing of a circus animal in black felt pen on white paper. Cut out the animal and glue to a sheet of coloured construction paper. Cut a large number of narrow strips of black construction paper for the students to use as bars on the circus cages. Glue the strips to the coloured construction paper. Link the circus cages together to form a circus train in the hallway. Cut out a large locomotive silhouette and mount in front of the circus train.

j) Brainstorm the words associated with animals. Select one word and make large cut-paper letters to form the word on a 24" x 18" sheet of construction

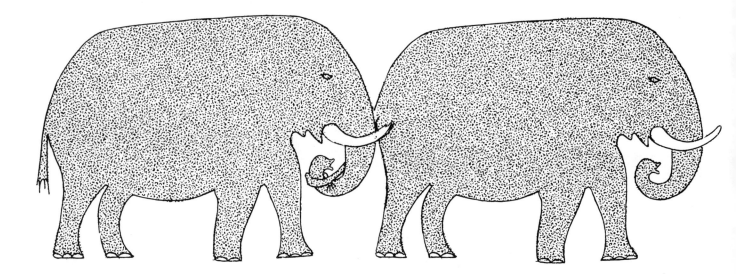

paper. Cut small silhouettes of the corresponding animal from a sheet of folded paper. Arrange the cut-paper silhouettes as a repeating pattern around the edge of the background sheet.

k) Draw for the class, one large animal on a length of mural paper. Arrange for a group of students to paint the animal in realistic or fantasy colours. Cut out the animal and staple to a hallway wall. Using cut-paper, have each student makes a seated figure in collage. Staple the cut-paper figures to the back of the giant animal. Possible choices of animals might be:

- crocodile
- hippopotamus
- dinosaur
- dragon
- rhinoceros
- camel
- kangaroo

Add reins, figures climbing up the animal on ladders, and so on, to complete the picture.

l) Hold a 'Favourite Animal Toy Day'. Students bring their favourite stuffed toy animal to school. Possible projects using the toys might include:

- drawings of the animals arranged on sheets of paper with examples of the students' writing

- drawings cut out and glued to a large sheet of mural paper as the first step in developing a toy shop
- drawings further embellished to create an imaginary setting for the animal
- drawings with the addition of collage clothing

m) Create new creatures from components of a wide variety of animals. Using photographs, draw parts of animals and join them to make new ones. Invent new animal names and create fantasy settings in paint or crayon.

n) Drawings of 'myself' could be cut out and glued to a piece of coloured construction paper. Then an imaginary setting which includes animals could be created. Some possibilities for the self portrait could include:
- me, the great explorer
- me, the animal trainer
- me, at the circus
- me, the pet shop owner

Combine the drawings with creative writing.

(2) Printmaking

a) On a sheet of 9" x 12" manila tag, make a large drawing of an animal. Cut out the drawing to create a stencil. Ink up the manila tag stencil with water-based black printer's ink and take a print on white duplicating paper. Cut out the print and glue to a larger sheet of white paper. Decorate the background with trees and flowers cut from brightly coloured tissue paper.

An extension of this project would be to print the animal a number of times on a long sheet of paper making sure to ink the stencil each time for dark prints. Lighter prints can be obtained by taking more than one print from each inking. By controlling the position of the prints and the density of inking, it is easy to develop a herd of animals where the lighter prints represent the animals in the distance. The project could remain this way or could be further enhanced through the addition of drawn or cut-paper images of plants, trees, mountains, and so on.

b) On an inked surface, such as glass, arborite or a sheet of manila tag taped to a desk, create a drawing of an animal by scraping away the ink with a blunt tool (such as a popsicle stick, handle of a paint brush). Texture the background by creating designs in the wet ink. Take a print on white duplicating paper. The project is easy to organize when students are able to come up to the printing table in turns.

c) Create animal prints from found objects. Add a small quantity of printer's ink or black paint to the object and print a number of times to create the animal shape. Develop a printed background from further found objects in full colour.

d) Create a cut-paper silhouette of a wild animal with a strong surface decoration, for example a tiger, giraffe, zebra, armadillo. Add scraps of paper to complete the decoration. Place the silhouette beneath a large sheet of newsprint and make a rubbing by using a piece of oil pastel on its side. Apply the crayon very lightly for the initial rubbings. Move the animal to a different position under the newsprint and increase the crayon pressure for the next rubbing. Each time the silhouette is moved, change the crayon pressure. Overlap the images each time to produce a group of images representing a herd.

e) Create a cut-paper silhouette of an animal in a story or nursery rhyme. Glue additional pieces of paper to the silhouette to create repeating stripes or dots. Ink up lightly in white printer's ink and take a print on black construction paper. Cut out the prints and mount them on a long sheet of brightly coloured mural paper.

f) From a length of mural paper, cut a big tree with a large number of branches. Staple the tree to a wall and from its branches suspend prints of monkeys. Use manila tag for the monkeys' shapes, and ink and print on white paper with black water-based printer's ink. Cut out the prints and staple into position on the tree. Cut-paper leaves in bright greens could be added later, together with a brightly coloured snake on the trunk.

g) With an X'Acto knife, cut printmaker's styrofoam into irregular shapes. Use the styrofoam imaginatively to create fantasy animals which fit into the styrofoam shapes provided. Make sure the outline drawing touches all sides. Decorate the surface of the animal with repeating patterns. Cut away the surplus styrofoam with an X'Acto knife or scissors. Ink up with black water-based printer's ink and print on white paper. Create colourful backgrounds with oil pastel on paint.

(3) Patternmaking and Design

a) Cut circles of white cardboard 10" in diameter. To this circle, have students glue a cut-paper image of an animal decorated in brightly coloured papers. Suspend all the circles at different heights from a wooden or plastic sports hoop as a class mobile. Glue the animals to each side of the white circles.

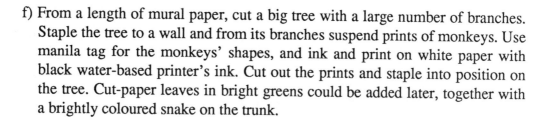

b) From scraps of styrofoam, make animal silhouettes to use for block printing. Set up a table for the printing and ink up the silhouettes in black water-based printer's ink. Print on long narrow pieces of white paper and arrange the prints as a repeating pattern. Hang the sheets of white paper vertically as banners throughout the classroom.

c) Cut an animal form from scrap textiles and glue to a square of felt with white glue. Decorate the animal with thread, buttons, sequins and other found objects. Glue the edges of the felt squares to a piece of burlap of contrasting colour. Better still, if parent volunteers are available, stitch the felt squares to the background. Exhibit the completed class project as a wall hanging. Small applique panels can also be made as individual projects.

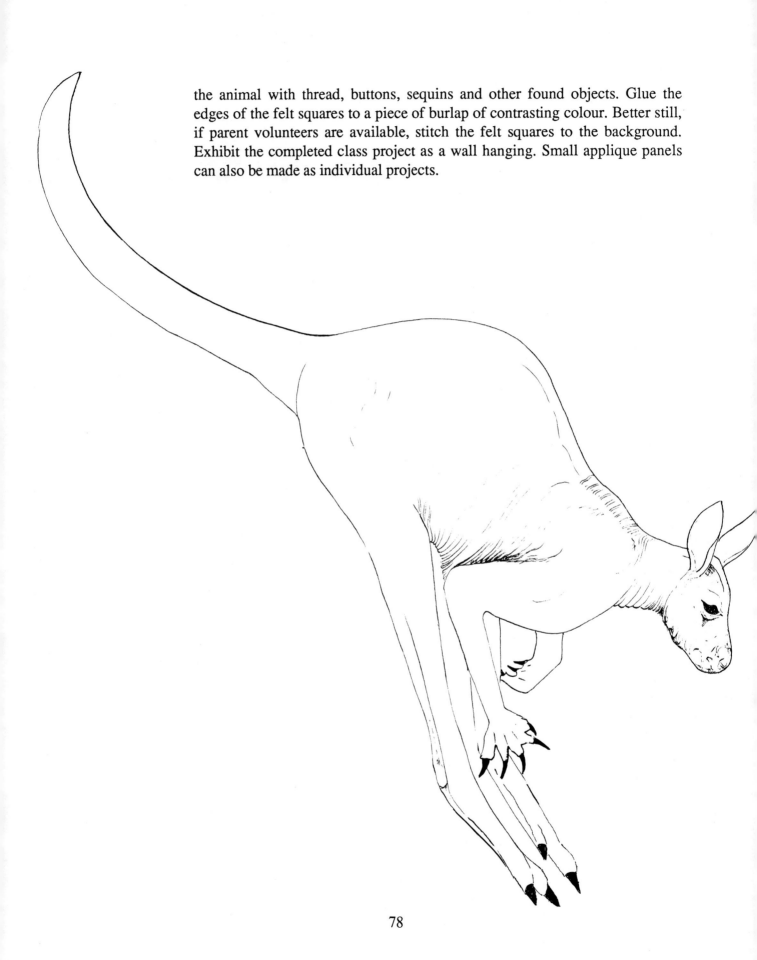

d) Cut black silhouettes of one animal in different sizes and positions. Exhibit them on a panel of brightly coloured paper, for example, black cats against a bright orange background at Hallowe'en.

e) Mount a decorated image of an animal on a sheet of coloured paper. Add a large decorated cut-paper letter to represent the initial letter of the animal.

f) Using images of birds and animals, create a class set of alphabet cards in paint, crayon, felt pen or collage, on squares of white cardboard.

B. THE IMAGE EXPRESSED IN THREE DIMENSIONS

(1) Ceramics

a) Decorate coiled or slab pots with an animal image. Build up the image by adding extra clay to the surface of the pot. Texture the surface of the animal so that it becomes quite pronounced on the pot.

b) Create small free-standing ceramic figures of animals. Choose animals with thick solid parts. Some examples are:
- rhinoceros
- hippopotamus
- elephant
- yak
- dinosaur
- buffalo

- cow
- bear

Make the body of the animal from two pinch-pots joined together. Later, when the body has hardened slightly, add additional clay to form legs, head, and so on. Texture the surface to simulate hair or decoration. Make a small hole in the body for air to escape and allow the ceramic animal to dry. Fire and exhibit as a class project in an imaginary background of cut-paper.

c) Roll out a slab of clay on a piece of plastic on folded newspaper. Draw an animal on a piece of manila tag and use the shape as a stencil. Draw around the animal stencil and cut away the excess clay leaving the animal silhouette. Decorate the surface of the animal by adding extra clay. Make a hole near the top edge so that the image can be used as a wall decoration or as part of a group mobile.

(2) Sculpture

a) Create fantasy animals from scrap materials by gluing interesting materials together.

b) Make papier mâché animals using crumpled paper, strips of manila tag or chicken wire for the armature. Paint with thick tempera and finish with Rhoplex or varnish.

c) Create animal glove puppets from scraps of felt. Relate the choice of animal to the children's reading programme.

ACKNOWLEDGMENTS

"Dogs and Cats and Bears and Bats" by Jack Prelutsky from *Random House Book of Poetry for Children.* ©1983 Random House. Reprinted by permission of Random House, Inc.

"First Things First" by Leland B. Jacobs from *Happiness Hill.* ©1960 Leland B. Jacobs. Published by Charles E. Merrill Books, Inc. Reprinted by permission of Leland B. Jacobs.

"I like Bugs" by Margaret Wise Brown from *The Fish With the Deep Sea Smile.* ©1965 Roberta B. Rauch. Published by E.P. Dutton. Reprinted with permission of Roberta Brown Rauch.

"Old Snake Has Gone to Sleep" by Margaret Brown from *Nibble, Nibble.* ©1987 Roberta B. Rauch. Published by W.R. Scott Publishers. Reprinted by permission of Roberta B. Rauch.

"A Pig Tale" by James Reeves from *The Blackbird in the Lilac.* ©1959 James Reeves. Reprinted with permission of E.P. Dutton & Co.

"Good Morning" by Muriel Sipe from *Sung Under the Silver Umbrella.* ©1935, Muriel Sipe. Printed by MacMillan Company. Used by permission of Mindy Koyanis for the author.

BIBLIOGRAPHY

Aliki, *At Mary Bloom's.* New York: Greenwillow Books, 1976.

Barchas, Sarah E., *I was Walking Down the Road.* New York: Scholastic Book Services, 1975.

Brown, Margaret Wise, *The Important Book,* New York: Harper & Row, 1949.

Brown, Margaret Wise, *Wheel on the Chimney,* New York: Harper & Row, 1954, 1987.

Carle, Eric, *The Very Hungry Caterpillar,* New York: G.P. Putnam's Sons, 1976.

Cleaver, Elizabeth, *The Loon's Necklace.* Toronto & New York: Oxford University Press, 1979.

Cook, Harold E., Shaker Music — A Manifestation of American Folk Culture. Lewisburg, PA: Bucknell University Press, 1973.

de Paolo, Tomie, *The Legend of the Bluebonnet.* New York: G.P. Putnam's Sons, 1983.

de Regniers, Beatrice Schenk, *Going for a Walk.* New York: Harper & Row, 1961.

Editors of Owl Magazine, *The Kids' Cat Book.* Toronto: Golden Book, Western Publishing, 1984.

_____ *The Kids' Dog Book.* Toronto: Golden Book, Western Publishing, 1984.

Fisher, Aileen, *Anybody Home?* New York: Thomas Y. Crowell, 1980.

_____ *But Ostriches.* New York: Thomas Y. Crowell, 1970.

Freedman, Russell, *Sharks.* New York: Holiday House, 1985.

Ginsburg, Mirra, *How the Sun Was Brought Back to the Sky.* New York: Macmillan Publishing Co., 1975.

Goble, Paul, *The Girl Who Loved Wild Horses.* Scarsdale, NY: Bradbury Press, 1978.

Harris, Lorele K., *Biography of a Mountain Gorilla.* New York: G.P. Putnam's Sons, 1981.

Heller, Ruth, *Animals Born Alive and Well.* New York: Grosset & Dunlap, 1982.

_____ *Chickens Aren't the Only Ones.* New York: Putnam Publishing Group, 1981.

Hutchins, Pat, *Good-night Owl.* New York: Macmillan Publishing Co., 1972.

_____ *One Hunter.* New York: Greenwillow Books, 1982.

_____ *Rosie's Walk.* New York: Macmillan Publishing Co., 1968.

Jarrell, Randall, *The Bat Poet.* New York: Macmillan Publishing Co., 1963.

Kalan, Robert, *Blue Sea.* New York: Greenwillow Books, 1979.

Kessler, Ethel and Leonard, *Do Baby Bears Sit in Chairs?* What Animals Do. New York: Doubleday & Co., 1961.

King, Deborah, *Puffin.* New York: Lothrop, Lee & Shepherd, 1984.

Kipling, Rudyard, *Just So Stories.* London: Macmillan, 1902; New York: New American Lib., paperback.

Kuskin, Karla, *The Rose on My Cake,* New York: Harper & Row, 1964.

Lane, Margaret, *The Frog, The Squirrel, The Beaver, The Fox, The Spider,* and *The Fish.* New York: Dial Press, 1981.

Longstaff, John, *Over in the Meadow.* New York: Harcourt, Brace & World, 1957.

Marshall, Alan, *People of the Dreamtime.* Melbourne, Australia: Hyland House, 1978.

Martin, Bill, *Brown Bear, Brown Bear, What Do You See?* New York: Holt, Rinehart & Winston, 1983.

_____ and Peggy Brogan, *Sounds of Numbers.* New York: Holt, Rinehart & Winston, 1972.

McClung, Robert M., *Gorilla.* New York: William Morrow and Co., 1984.

McCracken, Marlene J. And Robert A., *Myself.* Winnipeg, MB: Peguis Publishers, 1985, 1988.

McCracken, Robert A. and Marlene J., *Tiger Cub Readers.* Winnipeg, MB: Peguis Publishers, 1987.

Morris, William Barrett, *The Longest Journey in the World.* New York: Holt, Rinehart & Winston, 1970. (A Bill Martin Instant Reader).

Pearson, Susan, *Monday I Was an Alligator.* New York: Harper & Row, 1979.

Pienkowski, Jan, *Dinnertime.* Los Angeles, CA: Price, Stern & Sloan, 1981.

Prelutsky, Jack, *The Random House Book of Poetry for Children.* New York: Random House, 1983.

Provensen, Alice and Martin, illus., *A Peaceable Kingdom, The Shaker Abecedarius.* New York: Viking Press.

Reeves, James, *The Blackbird in the Lilac.* New York: E.P. dutton, 1959.

Roughsey, Dick, *The Rainbow Serpent.* Toronto & London: Wm. Collins & Sons, 1975.

Sipe, Muriel, *Sung Under the Silver Umbrella,* New York: The Macmillan Company, 1935.

Spier, Peter, *Gobble, Growl, Grunt.* New York: Scholastic Books, 1971.

_____ *Fox went Out on a Chilly Night.* New York: Doubleday, 1961

Sutton, Eve, *My Cat Likes to Hide in Boxes.* Toronto, New York: (Puffin) Penguin Books, 1973.

White, E.B., *Charlotte's Web,* New York: Harper & Row, 1952.

Wood, Audrey, *Quick as a Cricket.* Singapore: Child's Play International, 1982.

Wright, Freire, *Borrowed Feathers and Other Fables.* New York: Random House, 1977.